BIOTECHNOLOGICAL APPLICATIONS OF PLANT CULTURES

A CRC Series of

Current Topics in Plant Molecular Biology

Peter M. Gresshoff, Editor

Published Titles

Peter M. Gresshoff, PLANT GENOME ANALYSIS, 1994

Peter M. Gresshoff, PLANT RESPONSES TO THE ENVIRONMENT, 1993

Peter M. Gresshoff, PLANT BIOTECHNOLOGY AND DEVELOPMENT, 1992

Forthcoming Title

Peter M. Gresshoff, TECHNOLOGY TRANSFER OF PLANT BIOTECHNOLOGY

BIOTECHNOLOGICAL APPLICATIONS OF PLANT CULTURES

Edited by
Peter D. Shargool, Ph.D.
University of Saskatchewan
Saskatoon, Saskatchewan
Canada

That T. Ngo, Ph.D.
StressGen Biotechnologies Corporation
Victoria, British Columbia
Canada

CRC Press
Boca Raton Ann Arbor London Tokyo

Library of Congress Cataloging-in-Publication Data

Biotechnological applications of plant cultures / edited by Peter D. Shargool, That T. Ngo
p. cm. — (CRC series of current topics in plant molecular biology)
Includes bibliographical references and index.
ISBN 0-8493-8262-9
1. Plant biotechnology. 2. Plant cell culture. I. Shargool, Peter D. II. Ngo, T. T. (That Tjien), 1944- . III. Series.
TP248.27.P55B585 1994
631.5′3—dc20 94-8432
for Library of Congress CIP

Direct all inquiries to CRC Press, Inc., 2000 Corporate Blvd., N.W., Boca Raton, Florida 33431.

International Standard Book Number 0-8493-8262-9
Library of Congress Card Number 94-8432

Printed in the United States of America 1 2 3 4 5 6 7 8 9 0
Printed on acid-free paper

Preface

The plant kingdom continues to be a rich reservoir of pharmaceuticals for medicine. It has provided us with a variety of drugs from the most common one, the analgesic acetylsalicylate, to the more recently discovered powerful anticancer drug, Taxol. Continuous efforts to search for new drugs by screening the pharmacological properties of plant extracts obtained from the forests of the world have intensified.

Plant culture techniques play a pivotal role in providing a constant and unlimited supply of uniform and reproducible materials for experimentation. Furthermore, plant cultures are amenable to transformations by a variety of methods. Culture conditions can be manipulated to coax the cells to overproduce some chemicals, or even to produce chemicals that would not normally be made.

In addition to its importance in the discovery of new medicines, plant culture technology plays an even more important role in solving world hunger by the development of agricultural crops that give higher yields and are more resistant to pathogens and adverse environmental and climatic conditions. By successful expression of foreign antigens in plants, the idea of "edible vaccines" becomes a step closer to reality.

This volume was organized to present state-of-the-art reviews on current techniques in plant culture work. It covers four broad areas: (1) production of secondary metabolites by plant cells (Chapters 1 and 2); (2) plant cell transformation techniques (Chapters 3 and 4); (3) breeding and micropropagation techniques (Chapters 5 to 7); and (4) plant cell and tissue bioreactor design (Chapters 8 and 9).

We would like to thank all the authors, who are experts in their respective fields, for their important contributions. We also thank Mr. Jeff Holtmeier for his guidance and patience. The continuous and cheerful support of our families has made the completion of this book possible.

Peter D. Shargool
That T. Ngo

The Editors

Peter D. Shargool, Ph.D., is a full professor in the Department of Biochemistry at the University of Saskatchewan, Saskatoon, Saskatchewan, Canada. He graduated in 1962 from Birkbeck College, University of London, England, with a B.Sc. degree in Botany, Zoology, and Chemistry. Dr. Shargool obtained his M.Sc. degree in Biochemistry in 1965 and his Ph.D. in 1968 from the University of Alberta. In 1968, he became Assistant Professor at the University of Saskatchewan and in 1979 reached the rank of full professor. He has previously written two review articles on the subject of biotechnological applications of cultivated plant cells.

That T. Ngo, Ph.D., is Chairman and Chief Executive Officer of StressGen Biotechnologies Corporation in Victoria, British Columbia, and an Adjunct Research Biologist at the University of California, Irvine. Dr. Ngo graduated in 1970 from the College of Arts and Science, University of Saskatchewan, Saskatoon, Saskatchewan, Canada, with a B.Sc. in Biochemistry and obtained his Ph.D. degree in Biochemistry in 1974 from the same university. Dr. Ngo is a member of the American Association for the Advancement of Science, the American Chemical Society, the New York Academy of Sciences, and Sigma Xi and is a Fellow of the American Institute of Chemists. Dr. Ngo has presented over 40 invited lectures at national and international meetings, has published more than 130 research papers, and has edited four books in biotechnology. He is currently serving on the editorial board of three international scientific journals. His major research interests include downstream bioprocessing, affinity chromatography, molecular diagnostics, and plant biotechnology.

Table of Contents

Chapter 1
Ginseng Production in Cultures of *Panax ginseng* Cells 1
Tsutomu Furuya and Keiichi Ushiyama

Chapter 2
Increasing Secondary Metabolite Production in Plant Cell Cultures
with Fungal Elicitors 23
Ho Nam Chang and Sang Jun Sim

Chapter 3
Transformation of Plant Cells by Bombardment
with Microprojectiles 37
R. N. Chibbar and K. K. Kartha

Chapter 4
The Transformation of Legumes Using *Agrobacterium tumefaciens* 61
Mark C. Jordan and Shaun L. A. Hobbs

Chapter 5
Biotechnology Applications of Haploids 77
Alison M. R. Ferrie, C. E. Palmer, and Wilfred A. Keller

Chapter 6
Micropropagation 111
Kenneth L. Giles and Kenneth R. D. Friesen

Chapter 7
Artificial Seeds: A Comparison of Desiccation Tolerance
in Zygotic and Somatic Embryos 129
Bryan D. McKersie and Susan D. N. Van Acker

Chapter 8
Reactor Design for Plant Cell Suspension Culture 151
Gurmeet Singh and Wayne R. Curtis

Chapter 9
Reactor Design for Plant Root Culture 185
Gurmeet Singh and Wayne R. Curtis

Index 207

Ginseng Production in Cultures of *Panax ginseng* Cells

Tsutomu Furuya[1] and Keiichi Ushiyama[2]

[1]*Faculty of Science, Okayama University of Science, Okayama, Japan*

[2]*Nitto Denko Corporation, Ibaraki, Japan*

Introduction

The crude drug *ginseng* (called *ninjin* in Japanese), prepared from the roots of *Panax ginseng* C. A. Meyer (Araliaceae), is an important Oriental drug which stimulates the diminished metabolism of feeble patients. Many chemical, biochemical, and pharmacological investigations of ginseng have been carried out. In addition, plant biotechnological uses of ginseng have been widely reported. Recent advances in large scale culture of ginseng cells, the pharmacology of tissue-cultured ginseng, embryogenesis, and statistics applied to plantlet regeneration will be described in this chapter.

Large Scale Culture

Large scale culture of plant cells has developed the need for commercial production of secondary metabolites. The growth of plant cells in suspension is slow, and high inoculation density is required for rapid growth. The large size of plant cells, associated with a rigid cell wall and large vacuole, makes them sensitive to shear. The mixing, sampling, and aeration ratio in a bioreactor becomes a significant factor of growth and secondary product formation. Because of this, careful planning in scale-up is required.

0-8493-8262-9/94/$0.00+$.50

Table 1 *Content of main ginsenosides in tissue-cultured and cultivated ginseng*

Sample	Main ginsenosides (mg/g)							
	Rb_1	Rb_2	Rc	Rd	Rf	Re	Rg_1	Total
Chinese cultivated ginseng	2.1	1.1	1.5	0.8	1.4	2.3	2.8	12.0
Korean cultivated ginseng	1.8	0.8	1.2	0.8	1.3	2.7	2.1	10.7
Japanese cultivated ginseng	1.7	1.1	1.1	0.6	1.8	1.3	3.1	10.7
Tissue-cultured ginseng								
Lot. 21704	2.5	0.6	0.5	1.2	0.6	0.3	2.3	8.0
22707	2.8	0.5	0.4	1.5	0.7	0.3	2.0	8.2
23802	3.0	0.4	0.4	1.6	0.7	0.3	2.9	9.3
30806	3.1	0.5	0.4	1.2	0.6	0.4	1.9	8.1
01809	2.4	0.8	0.7	1.2	1.2	0.7	3.7	10.7
07903	1.8	0.4	0.5	0.8	1.0	0.6	3.0	8.1
06909	1.6	0.3	0.3	0.5	1.2	0.5	3.1	7.5

Note: Quantitative analysis of ginsenosides was carried out by HPLC method.

Conditions apparatus: 840 type HPLC station (Waters)
Column: TSK gel ODS80TM (4.6 mm × 250 mm)
Solvent: CH_3CN-H_2O (3:7) for ginsenoside Rb_1, Rb_2, Rc, Rd, Rf
CH_3CN-H_2O (19:81) for ginsenoside Rg_1, Re
Flow rate: 2.0 ml/min
Detection: UV (203 nm).

Keeping in mind the above-mentioned problems and the enormously fundamental studies described in the author's reviews,[1,2] the authors tried large scale production of ginseng cells with the support of Nitto Denko Corporation. That ginseng suspension culture in the scales of 2-ton and 20-ton tank bioreactors could be successfully achieved was demonstrated in 1982 and 1985. The productivity was more than 500 mg/l/d as dry material. The saponin ginsenosides content and the composition of the products were almost the same as that of cultivated ginseng[3] (Table 1).

The engineering of a new device for suspension culture in tank bioreactors of a scale of several tons was needed, and new ideas and technology were ultimately used. Moreover, investigations were made concerning repeated selection of cell lines, turbine type, agitation, valve improvement, and sterile growth conditions.

The results obtained from the above have shown that the increase of the growth ratio and dry weight is accompanied by an increase of saponin content. Thus, the mass production of ginseng cells has been carried out successfully for the first time in the world in a 20-ton tank bioreactor. The dried powder and extracts of tissue-cultured ginseng are nowadays widely used for health drinks, soups, cosmetics, and so on.

Pharmacology

For more than 2000 years ginseng has been used in traditional Oriental medicine as a tonic drug. Numerous reports of the biochemical and pharmacological effects of ginseng have appeared in the past 20 years. The recent studies on pharmacological activities will be profiled under the following five sections.

Hypoglycemic Activity

Among the pharmacological actions, Suzuki and Hikino[4,5] have been particularly interested in the hypoglycemic activity of ginseng and have isolated the glycans designated as panaxans A, B, C, D, and E as the hypoglycemic components. The effects of the main glycans, panaxans A and B, on the plasma insulin level, insulin secretion from pancreatic islets, insulin sensitivity, and insulin binding to isolate adipocytes have been confirmed. Moreover, the effects of panaxans A and B on the activities of hepatic key enzymes participating in the carbohydrate metabolism also have been examined by them.

Recently, Konno et al.[3] have isolated the glycans — callusans A, B, C, D, and E from the aqueous extract of tissue-cultured ginseng, according to the flow sheet as shown in Figure 1.

From chemical and physical data (Table 2), the structures of callusans are presumed to be heteroglycans having rhamnose, arabinose, and galactose as the sugar portion and a peptide portion.

Callusans A, D, and E exhibited significant hypoglycemic activity in normal mice, almost the same activity as that of panaxans A and B (Table 3). Callusan E, a main component of callusans, showed remarkable hypoglycemic activity in alloxan-induced hyperglycemic mice (Table 4).

Effects on the Function of the Stomach and Small Intestine

Effects of tissue-cultured ginseng on the function of the stomach and small intestine were compared with those of cultivated ginseng by the authors.[6] Ethanol extracts (50%) of the tissue-cultured and cultivated ginseng were administered in doses of 0.4 to 2.0 g/kg, singly (Table 5) or repeatedly (Table 6) and confirmed that the promotion of gastrointestinal propulsion in mice was significantly dose dependent.

The tissue-cultured ginseng also inhibited ulcer formation (induced by water immersion and restrained stress) and ligature of the pylorus. In contrast, the cultivated ginseng showed no such inhibitory action.

Gastric Secretion and Pepsin Activity

Effects of the tissue-cultured and cultivated ginseng on gastric secretion and pepsin activity were investigated by the authors.[7] Ethanol extracts (50%) of

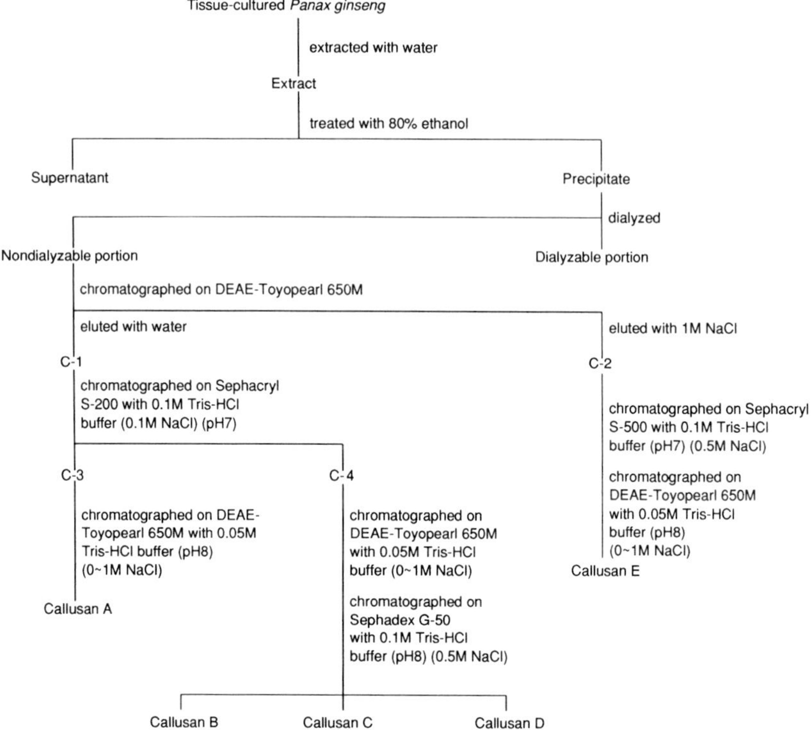

Figure 1. *Isolation of callusans A, B, C, D, and E from tissue-cultured* Panax ginseng.

both tissue-cultured and cultivated ginseng reduced gastric secretion (Table 7) and acid output (Table 8) in pylorus-ligated rats. They did not affect pepsin activity.

The tissue-cultured ginseng inhibited histamine- and pentagastrin-induced acid secretion in rats, whereas the cultivated ginseng showed no such effect. Tissue-cultured ginseng also suppressed acid secretion induced by 2-deoxy-D-glucose and baclofen [β-(*p*-chlorophenyl)-γ-aminobutyric acid], which are known to stimulate gastric acid secretion via the central nervous system (Table 9). However, they had no effect on acid secretion induced by vagal stimulation.

These results suggest that both tissue-cultured and cultivated ginseng may have an inhibitory effect on gastric secretion. The effect seems to be due to the inhibition of acid secretion via the central nervous system.

Effects on Blood Ethanol Concentration

In the course of pharmacological study of tissue-cultured and cultivated ginseng, the effects of 50% ethanol extracts of both materials on blood ethanol level in rats were investigated by the authors.[8] It has now been clarified that the extract of tissue-cultured ginseng reduced the blood ethanol level in 1 g/kg of

Table 2 Chemical and physical data of callusans A, B, C, D, and E

	Callusan A	Callusan B	Callusan C	Callusan D	Callusan E
Specific rotation	–22.9	–17.6	+30.4	+21.6	–28.8
Elemental analysis	C, 37.34 H, 6.41 N, 0.47	C, 33.84 H, 4.83 N, 1.34	C, 36.75 H, 5.90 N, 2.09	C, 35.22 H, 6.14 N, 3.19	C, 36.59 H, 6.31 N, 0.89
Infrared spectrum (cm^{-1})	3440, 1017	3450, 1060	3400, 1050	3400, 1041	3450, 1062
HNMR spectrum (δ)	4.91 (br) 5.16 (br) 5.34 (br)	4.96 (br) 5.21 (br) 5.40 (br)	4.94 (br) 5.19 (br) 5.36 (br)	5.12 (br) 5.26 (br)	5.15 (br) 5.35 (br)
Molecular weight (Mr)	9,100	14,000	5,600	1,950	580,000
Sugar contents (%)					
Phenol-H_2SO_4	61.9	38.1	64.9	41.1	42.4
Anthrone-H_2SO_4	37.4	28.1	40.8	30.9	35.0
Chromotropic acid-H_2SO_4	59.9	46.3	46.9	35.2	55.4
Carbazole-H_2SO_4	9.3	7.7	9.5	10.4	16.3
Peptide contents (%)					
Lowry method	0	0.8	13.0	19.2	8.6
Neutral sugar components (molar ratio)	Rhamnose, arabinose, xylose, galactose (0.2:0.8:1.4:1.0)	Arabinose, xylose (1.1:1.0)	Arabinose, galactose (0.3:1.0)	Arabinose, xylose, mannose, galactose, glucose (1.3:0.3:0.4:1.0:1.0)	Rhamnose, arabinose, xylose, galactose (0.1:0.7:0.1:1.0)

Table 3 *Effects of nondialyzable portion of tissue-cultured ginseng extract and callusans A, B, C, D, and E on plasma glucose level in normal mice*

		Relative glucose level				
	Dose	0	7		24(h[a])	
Drug	**(mg/kg, i.p.)**	**m[b]**	**m ± SE**	**%**	**m ± SE**	**%**
Control	—	100	106 ± 7	100	108 ± 7	100
Nondialyzable portion	10[4c]	100	51 ± 3**	48	74 ± 4**	69
Control	—	100	109 ± 3	100	98 ± 6	100
Callusan A	10	100	112 ± 3	103	94 ± 4	96
	30	100	90 ± 6	83	92 ± 3	94
	100	100	70 ± 6**	64	83 ± 3	85
Control	—	100	110 ± 6	100	111 ± 4	100
Callusan B	10	100	123 ± 7	112	134 ± 6	121
	30	100	104 ± 7	95	116 ± 4	105
	100	100	97 ± 4	88	104 ± 6	94
Control	—	100	107 ± 3	100	100 ± 3	100
Callusan C	10	100	108 ± 6	101	103 ± 5	103
	30	100	117 ± 5	109	113 ± 2	113
	100	100	110 ± 9	103	102 ± 6	102
Callusan D	10	100	108 ± 6	101	110 ± 9	110
	30	100	79 ± 5**	74	99 ± 6	99
	100	100	72 ± 5**	67	80 ± 3**	80
Control	—	100	111 ± 3	100	82 ± 5	100
Callusan E	10	100	77 ± 2**	69	83 ± 3	101
	30	100	61 ± 4**	55	71 ± 2	87
	100	100	64 ± 5**	58	75 ± 7	91

Note: n = 5; significantly different from the control $*p < 0.05$ or $**p < 0.01$.

[a] Time after administration.

[b] Plasma glucose level at 0 h and 140 to 170 mg/dl.

[c] Crude drug equivalent.

oral dose, but did not reduce that in 0.1 g/kg (Figure 2), and the effect of the tissue-cultured ginseng is slightly stronger than that of cultivated ginseng (Figure 3).

From the above experimental results, it is expected that tissue-cultured ginseng, as well as cultivated ginseng, may reduce the acute toxicity of ethanol and prevent ethanol-induced hepatic damage.

Effects on Peripheral Blood Flow and Blood Pressure

Effects of tissue-cultured ginseng on femoral and dermal blood flow and blood pressure, together with heartbeat rate were compared with those of cultivated ginseng by the authors.[9]

Table 4 *Effects of callusan E on plasma glucose level in alloxane-induced hyperglycemic mice*

		Relative glucose level				
	Dose	**0 (h)**	**7 (h)**		**24 (h)**	
Drug	**(mg/kg, i.p.)**	**m**	**m ± SE**	**%**	**m ± SE**	**%**
Control	—	100	105 ± 5	100	96 ± 8	100
Callusan E	10	100	97 ± 12	92	99 ± 7	103
	30	100	84 ± 17	80	106 ± 7	110
	100	100	53 ± 7**	50	104 ± 7	108

Note: n = 5; time is hours after administration; plasma glucose level at 0 h was 250 to 450 mg/dl. Significantly different from the control $*p < 0.05$ or $**p < 0.01$.

Table 5 *Effect of a single administration of tissue-cultured ginseng or cultivated ginseng on gastrointestinal propulsion in mice*

Drug	**Dose (mg/kg, p.o.)**	**No. of mice**	**Traversed[a] (%)**	**Promotion (%)**
Control	—	10	38.2 ± 2.2	—
Tissue-cultured	400	10	44.9 ± 2.1[e]	+17.5
ginseng[c]	800	10	48.2 ± 2.4[f]	+26.2
	2000	10	52.1 ± 4.1[f]	+36.4
Cultivated ginseng	400	10	38.1 ± 3.4	–0.3
(China, A)	800	10	36.7 ± 3.2	–3.9
	2000	10	41.7 ± 3.5	+9.2
Cultivated ginseng	400	10	40.6 ± 1.7	+6.3
(Korea)[b]	800	10	42.5 ± 2.6	+11.3
	2000	10	44.2 ± 2.7[d]	+15.7
Cultivated ginseng	400	10	40.5 ± 2.9	+ 6.0
(Japan)[b]	800	10	43.8 ± 2.3[d]	+14.7
	2000	10	44.3 ± 3.4	+16.0
Carbachol	0.3 (s.c.)	10	55.8 ± 4.4[f]	+46.1

[a] Each value represents the mean ± S.E.M. Significant dose-dependent promotion was observed: [b] $p < 0.10$, [c] $p < 0.01$. Significantly different from control: [d] $p < 0.10$, [e] $p < 0.05$, [f] $p < 0.01$.

Intravenous injections of 50% ethanol extracts of both types of ginseng were used to anesthetize rats for the test. The femoral blood flow of tissue-cultured and cultivated ginseng increased by 39.1 and 21 to 23%, respectively, with a dosage of 30 mg/kg (Figure 4).

Table 6 *Effect of repeated administrations of tissue-cultured ginseng or cultivated ginseng on gastrointestinal propulsion in mice*

Drug	Dose (mg/kg, p.o.)	No. of mice	Traversed[a] (%)	Promotion (%)
Control	—	10	34.9 ± 2.3	—
Tissue-cultured	400	10	36.6 ± 2.4	+ 4.9
ginseng[d]	800	10	48.1 ± 2.6[g]	+37.8
	2000	9	50.5 ± 4.0[g]	+44.7
Cultivated ginseng	400	9	43.5 ± 3.8[e]	+24.6
(China, A)	800	9	42.3 ± 5.0	+21.2
	2000	10	38.9 ± 2.3	+11.5
Cultivated ginseng	400	10	43.3 ± 3.0[f]	+24.1
(Korea)[c]	800	10	39.5 ± 2.7	+13.2
	2000	9	46.0 ± 3.0[g]	+31.8
Cultivated ginseng	400	10	43.0 ± 2.7[f]	+23.2
(Japan)[b]	800	10	44.9 ± 1.6[g]	+28.7
	2000	10	40.8 ± 2.4[e]	+16.9
Carbachol	0.3 (s.c.)	10	55.5 ± 2.3[g]	+59.0

[a] Each value represents the mean ± S.E.M. Significant dose-dependent promotion was observed: [b] $p < 0.10$, [c] $p < 0.05$, [d] $p < 0.01$. Significantly different from control: [e] $p < 0.10$, [f] $p < 0.05$, [g] $p < 0.01$.

Table 7 *Effect of tissue-cultured ginseng and cultivated ginseng on gastric secretion in pylorus-ligated rats*

Drug	Dose (mg/kg, i.d.)	No. of rats	Volume[a] (ml)	Inhibition (%)
Control	—	5	4.6 ± 1.1	—
Tissue-cultured	400	5	3.5 ± 0.4	+23.9
ginseng[b]	800	5	2.9 ± 0.5	+43.5
	2000	5	1.9 ± 0.4[c]	+58.7
Cultivated ginseng[b]	400	5	5.1 ± 0.7	–10.9
	800	5	3.3 ± 0.6	+28.3
	2000	5	3.1 ± 0.5	+32.6
Atropine sulfate	10	5	0.7 ± 0.3[d]	+84.8

[a] Each value represents the mean ± S.E.M. Significant dose-dependent inhibition was observed: [b] $p < 0.05$. Significantly different from control: [c] $p < 0.05$, [d] $p < 0.01$.

Table 8 *Effect of tissue-cultured ginseng and cultivated ginseng on gastric acid secretion in pylorus-ligated rats*

Drug	Dose (mg/kg, i.d.)	No. of rats	pH[a]	Acid output[a] (μeq/6 h)	Inhibition (%)
Control	—	5	2.01 ± 0.33	331.9 ± 117.1	—
Tissue-cultured ginseng	2000	5	3.51 ± 0.25[d]	61.8 ± 13.3[c]	+81.4
Cultivated ginseng	2000	5	3.01 ± 0.21[c]	95.5 ± 25.7[b]	+71.2
Atropine sulfate	10	5	3.11 ± 0.45[b]	46.3 ± 12.0[c]	+86.1

[a] Each value represents the mean ± S.E.M. Significantly different from control: [b] $p < 0.10$, [c] $p < 0.05$, [d] $p < 0.01$.

The dermal blood flow of tissue-cultured ginseng showed 9.9 to 27.4% increase with a dose of 1 to 30 mg/kg. It is noted that Korean cultivated ginseng produced a similar increase of 6.9 to 19.1%, but Japanese and Chinese types have no effect at the same dosage (Figure 5).

The depression of blood pressure of the femoral artery was observed in the range of 7.9 to 28.9% at the dose of 1 to 30 mg/kg of tissue-cultured ginseng (Figure 6). In the case of cultivated ginseng, almost the same results were confirmed, i.e., 24 to 37% depression at 30 mg/kg.

From these experimental results, it was concluded that the tissue-cultured ginseng may improve peripheral blood flow and blood pressure.

Embryogenesis

The ginseng plant requires a very long time (5 to 7 years) to grow to a size when it can be harvested and its seeds obtained (Figure 7A), and these seeds subsequently usually show low germination rates. The embryoid culture technique looks promising in solving these problems. Some recently reported methods have described the formation of embryoids using a callus from a root or young flower bud (Figure 7B, closed arrows).

The author and co-workers have found that a moderately high temperature induces embryogenesis from stable multiple shoots prepared from ginseng (Figure 7, open arrow).[10] The results are summarized in Figure 8.

The optimal temperature and treatment time are 30 to 40°C and 12 to 24 h, respectively. A lower effective temperature needs a longer treatment time. The number of embryoids was counted in experiments using a high temperature treatment time of 12 h, and the number was divided by the weight of inoculum segments. The average and standard deviations of four experiments were 9.3

Table 9 *Effect of tissue-cultured ginseng and cultivated ginseng on gastric acid secretion induced by histamine, pentagastrin, carbachol, 2-deoxy-D-glucose, baclofen, and vagal stimulation in rats*

Stimulation	Drug[a]	Dose (mg/kg, i.d.)	No. of Rats	Gastric acid output[b] (μeq/2 h)	Inhibition (%)
Histamine	Control	—	5	364.4 ± 22.8	—
	Tissue-cultured ginseng	2000	5	229.2 ± 26.4[e]	+37.1
	Cultivated ginseng	2000	5	324.2 ± 69.5	+11.0
	Cimetidine	10 (i.p.)	3	120.3 ± 27.9[e]	+67.0
Pentagastrin	Control	—	5	102.6 ± 9.8	—
	Tissue-cultured ginseng	2000	5	67.8 ± 10.7[d]	+33.9
	Cultivated ginseng	2000	5	108.4 ± 18.7	–5.7
	Cimetidine	10 (i.p.)	3	27.3 ± 8.5[e]	+73.4
Carbachol	Control	—	5	315.4 ± 36.6	—
	Tissue-cultured ginseng	2000	5	295.0 ± 29.4	+6.5
	Cultivated ginseng	2000	5	305.2 ± 25.5	+3.3
	Atropine sulfate	10	3	13.0 ± 4.2[e]	+95.9
2-Deoxy-D-glucose	Control	—	5	175.4 ± 41.1	—
	Tissue-cultured ginseng	2000	5	23.2 ± 5.2[e]	+86.8
	Cultivated ginseng	2000	5	79.0 ± 27.3[c]	+55.0
	Atropine sulfate	10	3	20.7 ± 5.2[e]	+88.2
Baclofen	Control	—	5	130.0 ± 33.3	—
	Tissue-cultured ginseng	2000	5	27.6 ± 4.5[e]	+78.8
	Cultivated ginseng	2000	5	48.8 ± 19.1[d]	+62.5
Vagal stimulation	Control	—	5	221.0 ± 30.3	—
	Tissue-cultured ginseng	2000	5	238.2 ± 30.5	–7.8
	Cultivated ginseng	2000	5	229.8 ± 25.6	–4.0
	Atropine sulfate	10	3	27.0 ± 1.2[e]	+87.8

[a] Each drug, except cimetidine, was given 60 min before histamine (20 mg/kg, s.c.), pentagastrin (1.0 mg/kg, s.c.), carbachol (0.1 mg/kg, s.c.), 2-deoxy-D-glucose (200 mg/kg, s.c.), baclofen (1.0 mg/kg, s.c.) administration and vagal stimulation. Cimetidine was given 20 min before histamine and pentagastrin administration. [b] Each value represents the mean ± S.E.M. Significantly different from control: [c] $p < 0.10$, [d] $p < 0.05$, [e] $p < 0.01$.

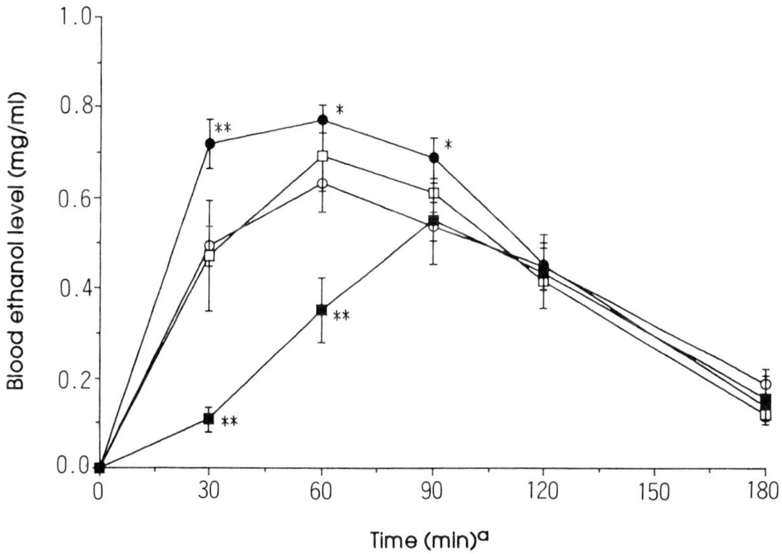

Figure 2. *Effects of tissue-cultured ginseng on blood ethanol level in rats. Data represent the means ± S.E.M. (n = 5).* [a] *= time after ethanol administration. *p < 0.05 and **p < 0.01 significantly different from control.* ○ *= control;* ● *= tissue-cultured ginseng 0.1 g/kg;* □ *= tissue-cultured ginseng 0.3 g/kg;* ■ *= tissue-cultured ginseng 1.0 g/kg.*

± 6.5 per g for 25°C, 75.7 ± 19.6 for 35°C, and 25.2 ± 12.1 for 43°C, respectively. The value for 25°C is at the level of spontaneous embryogenesis. At 35°C, the rate was 8 times. At 43°C, the number of embryoids formed was small, but the embryoids were easily separated from the other tissues since the growth of the other tissues was inhibited. These embryoids were formed on the surface of the differentiated tissue (Figure 9A). Heart-shaped embryoids were observed by using a microscope (Figure 9B). When these were cultured on a hormone-free MS medium for 6 weeks, young plantlets about 6 cm long were generated (Figure 10). It takes 3 months in total to get the plantlets after the treatment. This is half the time necessary to get plantlets of the same size from seeds (i.e., about 6 months).

Saponin components of these plantlets were examined by TLC and HPLC. Ginsenosides Rb_1 and Rg_1 and other saponins similar to those of cultivated ginseng were detected from these plantlets. The results suggest that high temperature treatment did not obstruct the biosynthesis of saponins. Thus, it appears that the high temperature treatment will be a useful method of stable mass propagation.

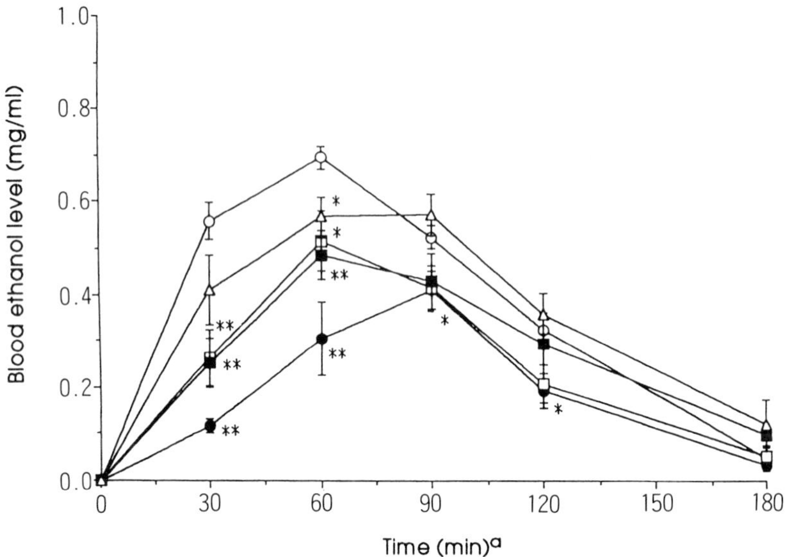

Figure 3. *Effects of various cultivated ginsengs on blood ethanol level in rats. Data represent the means ± S.E.M. (n = 5). [a] = time after ethanol administration (1.0 g/kg). *p < 0.05 and **p < 0.01 significantly different from control. ○ = control; ● = tissue-cultured ginseng; □ = Korean cultivated ginseng; ■ = Japanese cultivated ginseng; △ = Chinese cultivated ginseng.*

Statistical Methods for Plantlet Regeneration

The application of a statistic method to plantlet regeneration in ginseng callus culture was for the first time examined by the authors.[11] The shoots were induced from ginseng callus on a MS medium supplemented with kinetin (0.2 mg/l). On the basis of the results of L8 orthogonal array experiments, KNO_3 and $MgSO_4$ concentrations were chosen as significant factors in root formation (Tables 10 and 11). The effects of KNO_3 and $MgSO_4$ were then simulated by the Box-Wilson method (Table 12, Figure 11).

This study also demonstrated that a modified MS medium containing 185 mg/l $MgSO_4$ and no NH_4NO_3 or plant growth regulators is optimal for root formation from shoots. Plantlets formed under the above conditions are now growing well in greenhouses.

Conclusion

Part of the ginseng harvested in Japan is exported to Hong Kong and Korea, from which they are distributed to various Asian countries and even back to

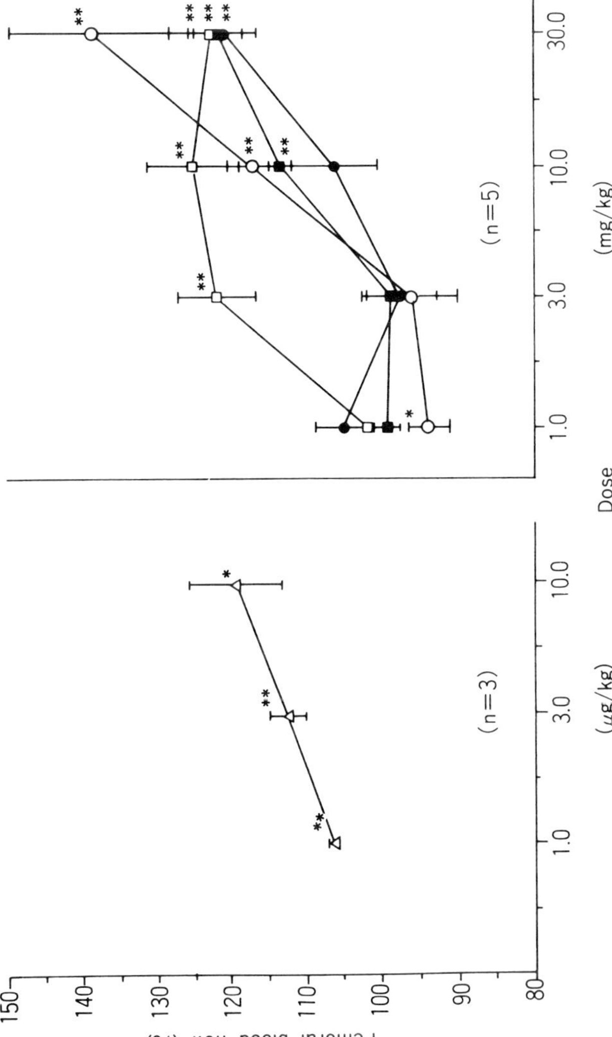

Figure 4. *Effects of tissue-cultured and cultivated ginsengs on blood flow of the femoral artery in rats. Data represent the means ± S.E.M.* $^{*}p < 0.05$ *and* $^{**}p < 0.01$ *significantly different from control.* △ = *phentolamine;* ○ = *tissue-cultured ginseng;* ● = *Korean cultivated ginseng;* □ = *Japanese cultivated ginseng;* ■ = *Chinese cultivated ginseng.*

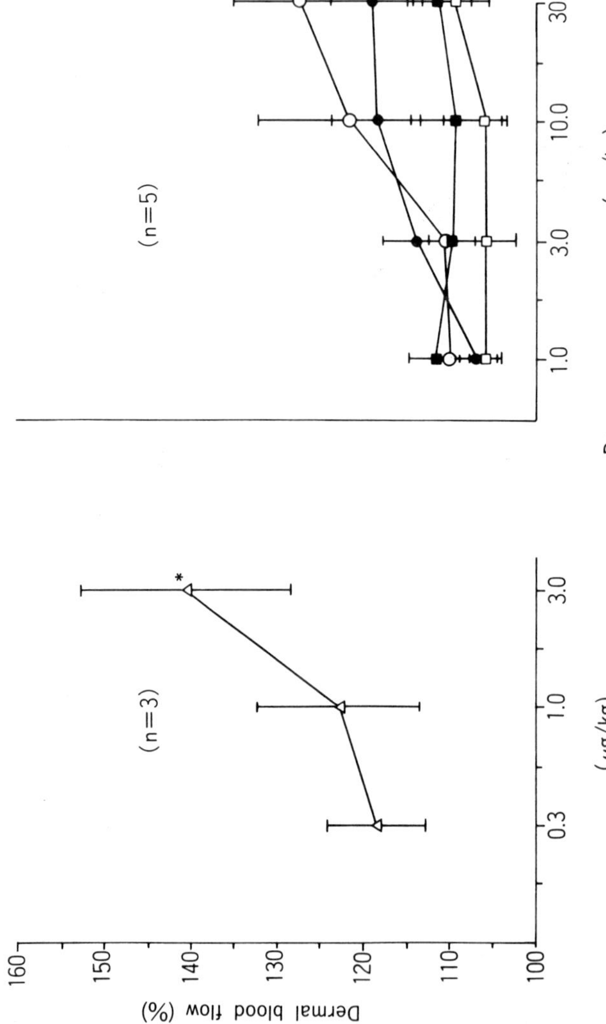

Figure 5. *Effects of tissue-cultured and cultivated ginsengs on blood flow of abdominal skin in rats. Data represent the means ± S.E.M. *p < 0.05 and **p < 0.01 significantly different from control.* △ *= papaverine;* ○ *= tissue-cultured ginseng;* ● *= Korean cultivated ginseng;* □ *= Japanese cultivated ginseng;* ■ *= Chinese cultivated ginseng.*

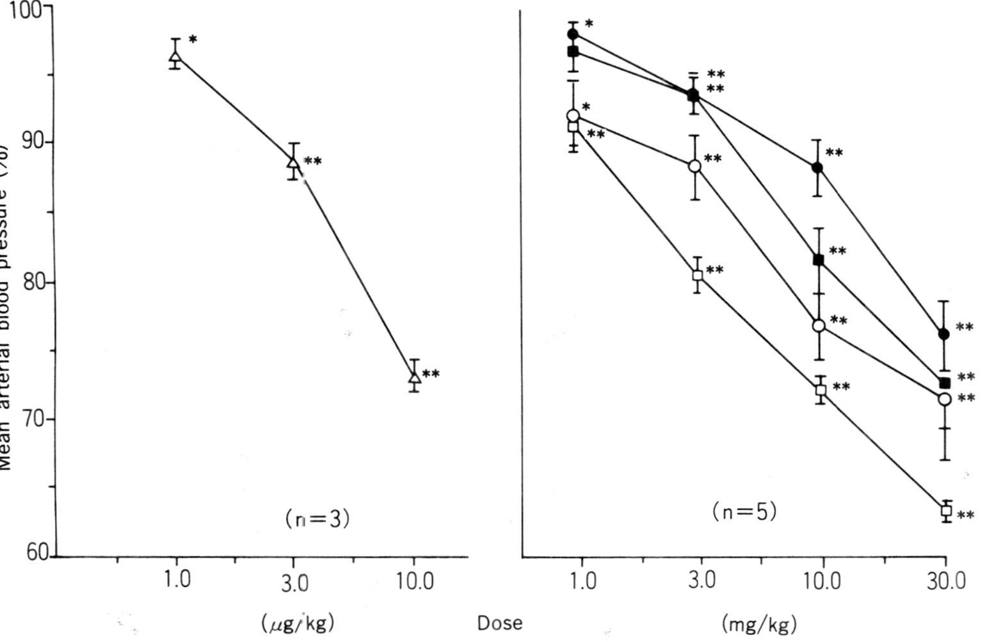

Figure 6. *Effects of tissue-cultured and cultivated ginsengs on blood pressure of the femoral artery in rats. Data represent the means ± S.E.M. *p < 0.05 and **p < 0.01 significantly different from control.* △ = *phentolamine;* ○ = *tissue-cultured ginseng;* ● = *Korean cultivated ginseng;* □ = *Japanese cultivated ginseng;* ■ = *Chinese cultivated ginseng.*

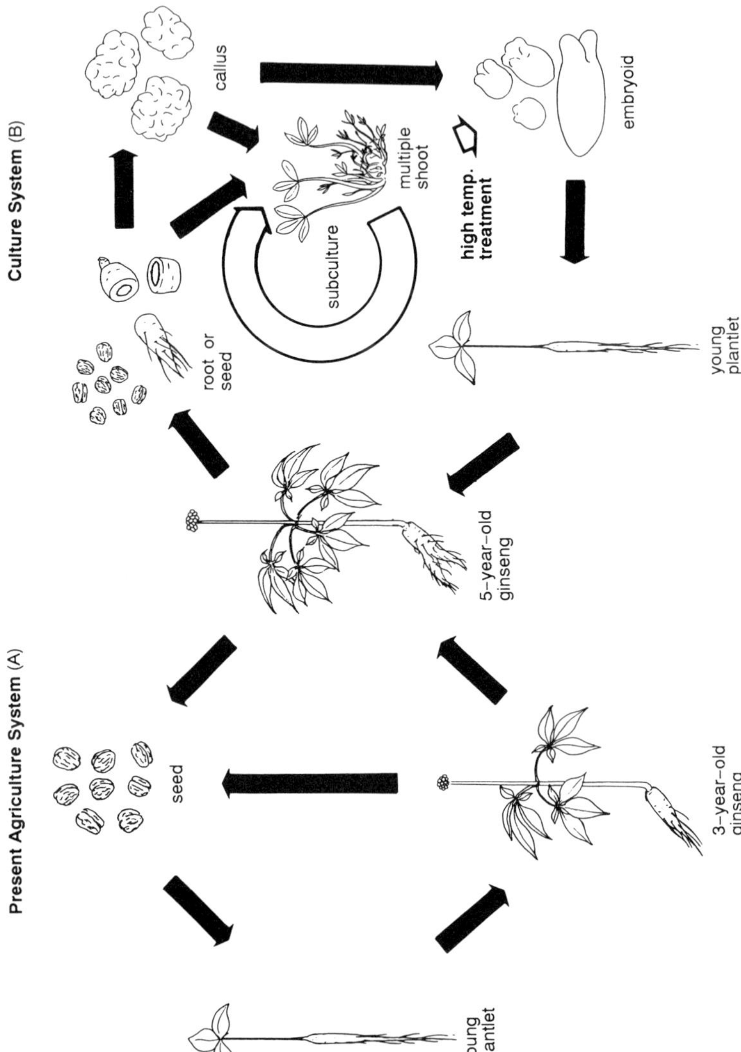

Figure 7. *Comparison of the present agriculture system (A) and the tissue culture system (B) of* Panax ginseng. *Present techniques are displayed by closed arrows, and findings are displayed by open arrows.*

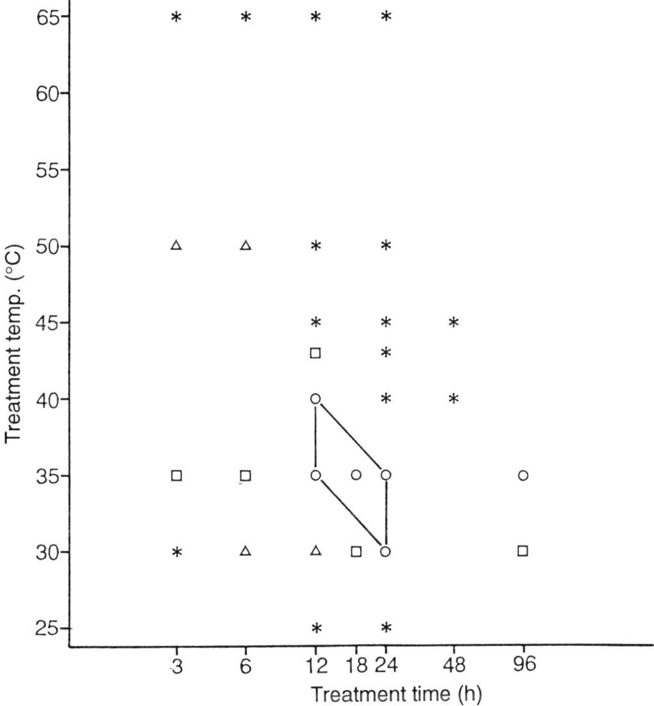

Figure 8. *Effective area of embryoids formation by high temperature treatment. Multiple shoot was obtained in 1978 by the method of Furuya et al. from 5-year-old ginseng root cultivated in Korea and subcultured at 20°C under illumination (5000 lux for 16 h/d). Each 5 to 6 segments of these redifferentiated tissues were transplanted on Murashige & Skoog (MS) media supplemented with kinetin (1 mg/l) and incubated at a high temperature for 3 to 96 h. After the treatment, the segments were cultured in a 12 to 25°C gradient cycle per day under illumination (5,000 to 10,000 lux gradient, 14 h/d) for 6 weeks, and embryoid-formed segments were counted. The frequency (%) was expressed as the number of embryoid-formed segments per number of transplanted segments (○ ≥ 65%; 65% > □ ≥ 50%; 50% > △ ≥ 30%; 30% > *). The data were obtained from three independent experiments.*

Japan. Ginseng is an official drug in Japan, Korea, China, the U.S., Switzerland, Austria, Germany, and Russia. In countries with high ginseng consumption, especially Japan, Korea, and the U.S., ginseng is recognized as a health food and drink.

Tissue-cultured ginseng produced by a 20-ton tank bioreactor is almost equal to the cultivated one in the quality and quantity of ginseng saponins and other components. Near-identical pharmacological actions were also demonstrated and no mutagenic potential was detected. Therefore, the large scale culture of ginseng is advantageous for supplying ginseng.

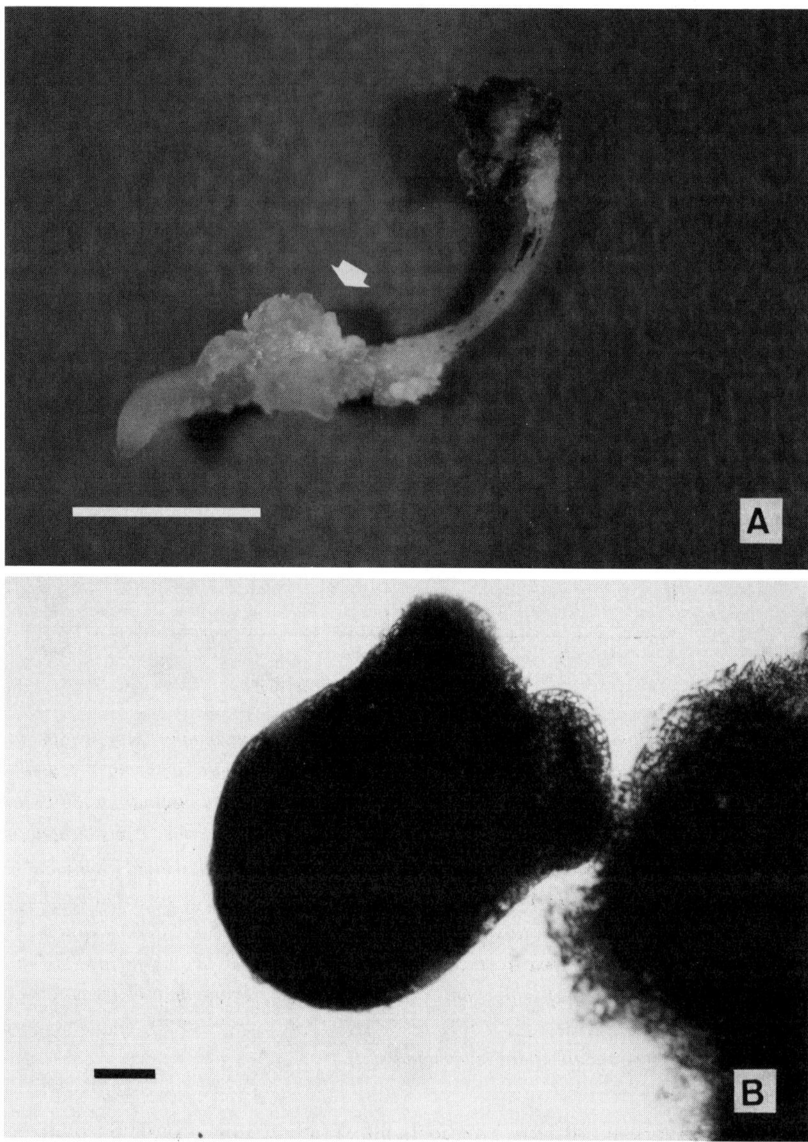

Figure 9. *Embryoids formed by high temperature treatment. (A) Embryoids formed on the surface of redifferentiated tissue. A multiple shoot tissue was treated at 35°C for 12 h and cultured for 6 weeks. The arrow shows embryoids formed on the surface of an adventitious root taken from the tissue. Scale bar = 1 cm. (B) Heart-shaped embryoid. It was separated from the surface of treated redifferentiated tissue. Scale bar = 100 μm.*

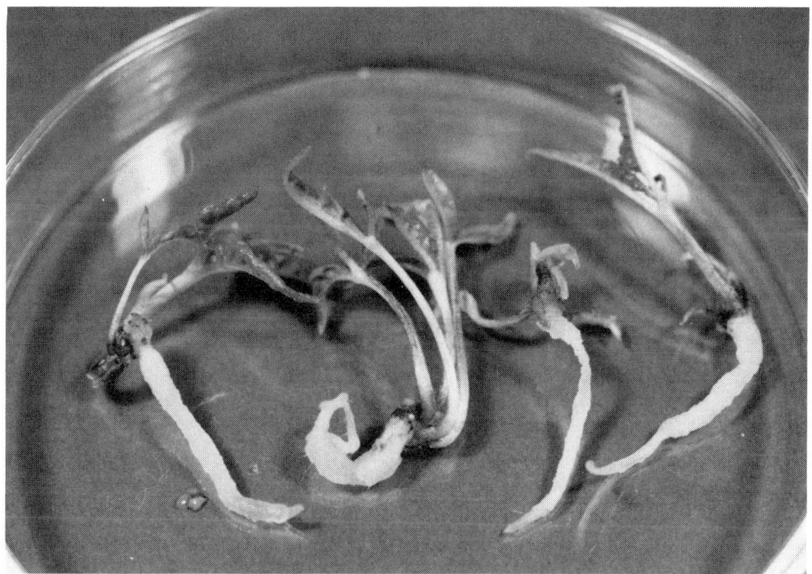

Figure 10. *Young plantlets regenerated from the embryoids. The embryoids were cultured on hormone-free MS media for 6 weeks.*

Table 10 *Effects of chemicals on root formation*

No.	Concentration (mg/l) NH_4NO_3	KNO_3	KH_2PO_4	$MgSO_4$	Root formation (%)
1	1650	1900	250	370	0
2	1650	1900	85	125	28
3	1650	125	250	125	33
4	1650	125	85	370	6
5	0	1900	250	125	33
6	0	1900	85	370	28
7	0	125	250	370	6
8	0	125	85	125	0

Table 11 *Analysis of variance on root formation*

No.	Factor	s.s.	d.f.	m.s.	Fo	Ratio (%)
1	NH_4NO_3	0	1	0	0	0
2	KNO_3	242	1	242	13.44	14.24
4	KH_2PO_4	12.50	1	12.50	0.69	0
7	$MgSO_4$	364.50	1	364.50	20.24	22.02
3	$NH_4NO_3 \times KNO_3$	544.50	1	544.50	30.24	33.46
6	$KNO_3 \times KH_2PO_4$	392.00	1	392.00	21.77	23.77
5	error	18.01	1	18.01		6.51
Total		1573.5	7			100.00

Note: s.s. = sum of square; d.f. = degree of freedom; m.s. = mean square; Fo = ratio of mean square.

Table 12 *Effects of KNO_3 and $MgSO_4$ on root formation*

No.	Concentration (mg/l)		Root formation (%)
	KNO_3	$MgSO_4$	
1	640	79	6
2	3420	79	39
3	640	327	28
4	3420	327	44
5	4000	150	44
6	100	150	56
7	2000	400	56
8	2000	50	28
9	2000	150	61
10	2000	150	78
11	2000	150	61
12	2000	150	61

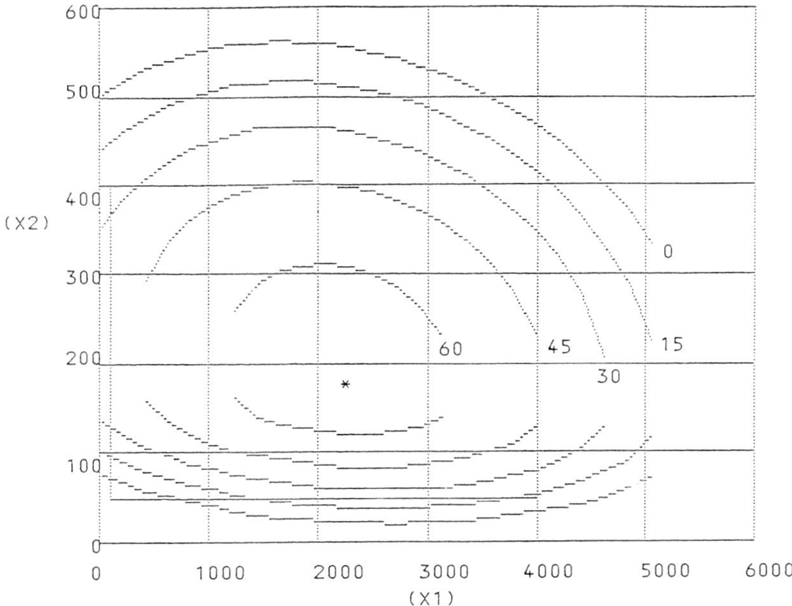

Figure 11. *Effects of KNO_3 (X1) and $MgSO_4$ (X2) on root formation. Numbers in the figure are the rates of root formation.*

References

1. **Furuya, T.,** Saponins (ginseng saponins), in *Cell Culture and Somatic Cell Genetics of Plants,* Vol. 5, Constabel, F. and Vasil, I. K., Eds., Academic Press, San Diego, 1988, chap. 12.

2. **Furuya, T.,** Production of useful compounds by plant cell cultures — *de novo* synthesis and biotransformation, *Yakugaku Zasshi,* 108, 675, 1988.

3. **Konno, C., Suzuki, Y., and Oshima, Y.,** Hypoglycemic activity of glycans of cultivated and tissue cultured *Panax ginseng* roots, *Nitto Tech. Rep.,* 28, 69, 1990.

4. **Suzuki, Y. and Hikino, H.,** Mechanisms of hypoglycemic activity of panaxans A and B, glycans of *Panax ginseng* roots: effects on the key enzymes of glucose metabolism in the liver of mice, *Phytother. Res.,* 3, 15, 1989.

5. **Suzuki, Y. and Hikino, H.,** Mechanisms of hypoglycemic activity of panaxans A and B, glycans of *Panax ginseng* roots: effects on plasma level, secretion, sensitivity and binding of insulin in mice, *Phytother. Res.,* 3, 20, 1989.

6. **Suzuki, Y., Ito, Y., Konno, C., and Furuya, T.,** Effects of tissue cultured ginseng on the function of the stomach and small intestine, *Yakugaku Zasshi,* 111, 765, 1991.

7. **Suzuki, Y., Ito, Y., Konno, C., and Furuya, T.,** Effects of tissue cultured ginseng on gastric secretion and pepsin activity, *Yakugaku Zasshi,* 111, 770, 1991.

8. **Suzuki, Y., Konno, C., and Furuya, T.,** Effects of tissue cultured ginseng on blood ethanol concentration in rats, *Nitto Tech. Rep.,* 30, 54, 1992.

9. **Suzuki, Y., Konno, C., and Furuya, T.,** Effects of tissue cultured ginseng on blood pressure and peripheral blood flow in rats, *Nitto Tech. Rep.,* 30, 60, 1992.

10. **Asaka, I., Ii, I., Yoshikawa, T., Hirotani, M., and Furuya, T.,** Embryoid formation by high temperature treatment from multiple shoots of *Panax ginseng*, *Planta Medica,* 59, 345, 1993.

11. **Furuya, T., Ohba, T., Oda, H., Miyamoto, Y., Higashitsuji, M., and Ushiyama, K.,** Application of statistical methods to plantlet regeneration in *Panax ginseng* callus culture, *Plant Tissue Cult. Lett.,* 6, 125, 1989.

Increasing Secondary Metabolite Production in Plant Cell Cultures with Fungal Elicitors

Ho Nam Chang and Sang Jun Sim

Bioprocess Research Engineering Center and Department of Chemical Engineering, Korea Advanced Institute of Science and Technology, Daeduk Science Town, Taejon, Korea

Introduction

In recent years, plant cell suspension cultures, considered to be viable alternatives to natural plant extraction, produce valuable biochemicals from plants and offer simplicity and potential economical benefits. Unfortunately, these benefits are hampered by low yields of the desired products, a lack of understanding of basic secondary metabolism in cultured plant cells, and difficulties in product separation. In order to improve the yield of product, various techniques have been tried including cell line selection, cell immobilization, manipulation of culture conditions (nutrients, pH, aeration), alteration of cell metabolism for increased production, and effective separation of desired products. In spite of intensive investigations, however, few processes based on plant cell culture have been developed on an industrial scale, except for the production of shikonin derivatives by *Lithospermum erythrorhizon*.[1]

0-8493-8262-9/94/$0.00+$.50

To some degree, the low yield of plant cell culture can be overcome by repeatedly screening for clones of high producing cells. However, the selection of high producing cultures requires tedious labor and often shows a decline in productivity upon serial propagation; therefore, scaling-up may be difficult. Elicitation of secondary metabolites on plant cell culture has lately received increasing attention, because elicitation of plant cells leads to increased yields and shortened fermentation time.[2]

In this chapter, we will discuss increasing secondary metabolites production with fungal elicitors and other stress factors (e.g., hormone levels, *in situ* extraction, etc.) that affect the secondary metabolism in plant cell culture.

Plant–Microbe Interaction

It is known that microbial invasions of normal plants lead to drastic changes in the secondary metabolism of plants. Plants have numerous types of defense mechanisms against this invasion by microorganisms. Plant diseases are caused by deleterious interactions among fungi, bacteria, or viruses. When plants are infected by a pathogen, a number of biochemical changes may occur that act cooperatively to stop further spread of the disease. One of the most common plant reactions in response to a pathogen invasion, brings about the precipitous death or hypersensitive reaction of those plant cells.[3–5] A major limitation in studying plant cell responses to microorganisms is the few defined elicitors available. A well-known defense mechanism is the accumulation at the infection site of phytoalexins, which are antimicrobial compounds of low molecular weight.[6] They are produced by plant tissues during infection, and the accumulation of phytoalexins can be induced by the molecules of either biotic or abiotic origin called *elicitors.* These elicitors are a special group of triggering factors that can be obtained from fungi, and crude preparations are easily screened for antibiotic phytochemical-inducing activity with cultured plant cells.

The biotic elicitors include glucan, glycoprotein, various fungal cell wall components, and endogenous materials derived by plant cell walls. Abiotic elicitors include UV irradiation, heavy metal ions, and detergents and are responsible for stimulating factors of secondary plant metabolism.[7] The biotic and abiotic elicitors have been used to increase the production, or to induce the *de novo* synthesis, of secondary metabolites in cell cultures.[2,8,9]

A considerable number of fungal and bacterial elicitors have been well-characterized, and several treatments have been published on the theory of recognition of pathogens by host plants on a molecular level.[10–12] However, little work has been performed in the regulatory mechanisms involved in their synthesis.

A plant-microbe interaction can be compatible or incompatible. It is dependent upon the genetic constitution of the two organisms. A nonhost microbe interaction is incompatible, by definition, and its resistance is a highly desired trait in economically valuable plants. The characteristics of many nonhost fungus interactions include limited hyphal growth and the accumulation of hypersensitive cell death. The spread of pathogens will be stopped by the accumulation of phytoalexin if an incompatible pathogen triggers a host response, which decreases the growth rate after pathogens. On the other hand, the compatible pathogen may be avoided by triggering the inhibitory response and, as a result, avoiding the toxic effects of the phytoalexin.[13] For example, under optimal conditions in inoculated plants, plant defense responses such as callose deposition, host cell necrosis, and phytoalexin accumulation are incompatible interactions, not compatible ones.[14] Table 1 shows the typical secondary metabolites elicited in plant cell cultures.

Action of Elicitors

The *Phytophthora megasperma* f. sp. *glycinea*-soybean system is useful for studying changes in metabolism-associated phytoalexin production. The β-glucan elicitor of the mycelial walls of the fungal pathogen, Pmg, may be one of the first elicitors discovered, and it is one of the most potent biotic elicitors.[5,13,14] The elicitor was found to be a carbohydrate structurally similar to portions of mycelial walls. It is known that a branched β-1,3-, β-1,6-linked heptaglucan of defined structure is the smallest fragment with high elicitor activity in soybeans.[10,36]

As mentioned above, plants have a variety of defense mechanisms against potential pathogens. Several of these mechanisms include phytoalexin production, reinforcement of plant cell walls through the deposition of callose, the synthesis of lignin-like materials, or the enhancement of the activities of certain hydrolytic enzymes such as chitinase or glucanase.[37,38] Soybean is susceptible to Pmg and shows race-specific resistance,[39] whereas parsley is not a host for this fungus.[40] Resistant soybean plants rapidly accumulate isoflavonoid phytoalexins in the infected tissue,[41] and parsley cells excrete coumarine derivatives into the infected droplets.[42]

Plant cells perceive appropriate signals against invasion by pathogens. The subsequent induction processes are limited to the invading area. Thus, the recognition may be dependent on the detection of biochemical features of the pathogen by plasma membrane components of the plant.[43] Schmidt and Ebel[44] demonstrated that the stimulation of phytoalexin formation by the elicitor is a membrane-mediated process and found that the elicitor binds to a receptor localized in the plasma membrane. The binding affinity appears to correspond to their efficiency in inducing glyceollin synthesis in soybean cotyledons which indicates that the binding site of the plant may be identical to the primary elicitor target.

Some of the changes in protein phosphorylation are among the earliest known events following elicitation. Rapid changes in the phosphorylation pattern of proteins were observed in elicitor-treated plants. Another investigation

Table 1 *Typical secondary metabolite elicited in plant cell culture*

Plant cells	Secondary metabolites	Elicitors	Ref.
Papaver somniferum	Sanguinarine	Fungal mycelia	5
	Morphine, cadeine	Fungal spores	16
Catharanthus roseus	Ajmalicine, catharanthine	Fungal homogenate	17
Lithospermum erythrorhizon	Shikonin	Agaropectin	18
		Endogeneous polysaccharide	19
		Fungal culture filtrate	20
Glycine max	Glyceollin	Chitosan, poly-L-lysine	21
		Fungal glucan	22
Phaseolus vulgaris	Kievitone, phaseollin	Fungal cell wall material	23
		Ribonuclease	24
		Fungal glycoprotein	25
		Fungal polysaccharide	26
Petroselinum hortense	Coumarins	Fungal polysaccharide	27
Petroselinum crispum	Furanocoumarins	Fungal cell wall materials	28
Ruta graveolens	Acidone expoxide	Chitosan	29
		Fungal polysaccharide	30
Thalictrum rugosum	Berberine	Yeast carbohydrate	31
Eschscholtzia californica	Sanguinarine	Fungal polysaccharides, yeast extracts	32
Datura stramonium	Lubimin	Fungal spores	33
Ricinus communis	Casbene	Fungal filtrate	34
Medicargo sativa	Pterocarpan	Fungal cell wall materials	35

has shown a dependency of Ca^{2+} of the plasma membrane necessary for 1,3-β-glucan synthase.[45] Furthermore, the possibility of K^+ efflux being added to the depolarization of the membrane has also been discussed.[46,47] These effects were correlated with the activation of the genes involved in the accumulation of phenylpropanoid derivatives.[48–50] Therefore, protein phosphorylation and the presence of Ca^{2+} in the medium play an important role in many signal transduction in plant cells.

Fungal Elicitation Combined with Other Stresses

Increased production of secondary metabolites may be induced by UV radiation, heat-shock treatment, osmotic stresses, and hormone deletion as well as elicitation. Radiation of dark-grown cell suspension cultures of parsley with UV light, under standard conditions, leads to the accumulation of flavonoid glycosides.[51] These compounds are putatively UV-protected. There are many challenges (such as UV-radiation and fungal pathogens) for plants. Biosynthesis of these two groups (abiotic and biotic elicitors) of plant defense mechanism proceeds via the pathway of general phenylpropanoid metabolism. Phenylalanine ammonia-lyase (PAL) and chalcone synthase (CHS) catalyze key reactions in the biosynthesis of the isoflavonoid phytoalexins. Two stimuli activate the same pathway through different mechanisms, UV light acting through a photoreceptor and an elicitor, possibly in part through protein phosphorylation processes.[49] While both treatments stimulate the general pathway, a maximal rate of enzyme synthesis is typically reached earlier with elicitor than with UV radiation.[52]

High temperature is another particularly important environmental stress factor. Heat shock induces the production of a number of specific proteins. Plants produce a specific set of heat shock proteins (HSPs) when tissue temperatures are increased, either gradually or abruptly.[53,54] The molecular mechanisms underlying the appearance of HSPs have been extensively studied in *Drosophila mellanogaster*. HSP expression is a major regulating point, the transcription activation of the HSP genes.[55,56] Changes in environmental or physiological conditions can alter the expression of a resistance response. Walter[57] described some characteristic features of the heat shock, UV light, and fungal elicitor response of parsley cell suspension cultures.[57] Stress responses in this system are expressed in the order of heat shock, fungal elicitor, and UV light.

Sanguinarine was first isolated from *Papaver* callus.[58] This compound was considered as a phytoalexin in tissue culture; moreover, this alkaloid possesses significant biological potency against incompatible pathogens, with the exception of the *Papaver* pathogen, *Dendryphion penillatum,* which displayed minimum susceptibility to sanguinarine.[59] This is due to its ability to metabolize the compound enhanced by the loss of the alkaloid from the medium, followed by subsequent accumulation and apparent disappearance from the mycelium. When hormone deletion was combined with elicitation, sanguinarine levels approached a level 50- to 500-fold greater than that of nonelicitated suspension growth in hormone medium. Furthermore, during the elicitation process, substantial amounts of sanguinarine accumulated in the medium.[60] In addition, by osmotic stresses, the content of secondary metabolites markedly increased. When fungal elicitation and osmotic stress treatment were sequentially applied to the cultures of *Catharanthus roseus,* the additive effect on alkaloids accumulation was three times higher than in cells cultured without treatment. In addition, the majority of the alkaloids was found in the medium (Figure 1).[61] From the above results, the secretion of the products may be enhanced through elicitation and other environmental stress factors.

Production and Secretion of Secondary Metabolites with Fungal Elicitation

The major disadvantage of immobilized cells is that most plant products are stored within cell vacuoles or cell walls. Efficient product release is essential to the performance of an immobilized cell reactor. Product release may be obtained through variations in pH, the use of permeabilizing agents,[62] and by *in situ* extraction. *In situ* extraction and adsorption can be performed by adding inert hydrophobic chemicals (liquid or solid) into the culture while the chemicals have high adsorption capacity for the hydrophobic plant products. These techniques have been practiced not only in microbial cell cultures but also in plant cell cultures.[63] Recently, it was reported that elicitation with fungal elicitor enhanced the secretion of product in plant cell cultures.

Shikonin derivatives are produced from cell suspension cultures of *Lithospermum erythrorhizon,* which has been used as a red pigment for coloring lipstick and in medicine for burn and hemorrhoid treatments. The fact that the biosynthesis of a certain acidic polysaccharide was necessary for the *Lithospermum* cells to initiate shikonin biosynthesis was demonstrated by Fukui et al.,[18] whose research has shown that shikonin-producing cells contain the inducement of biosynthesis of normal secondary metabolites of higher plants.[19]

The effect of elicitation with fungal elicitor on shikonin production was also studied.[20] From the contaminated cell culture of *L. erythrorhizon, penicillium* species were isolated, and the synergistic effects on shikonin production of *in situ* extraction by *n*-hexadecane and elicitation were studied. Both techniques resulted in a shikonin productivity increase 65 times higher than that of the control (Figure 2). Simultaneous uses of *in situ* extraction and elicitor treatment may be feasible for plant-derived secondary metabolites production that is normally stored intracellularly in vacuoles of plant cells.

Byun et al.[32] showed another example of the simultaneous use of both techniques. Large increases in sanguinarine production were observed in suspension-cultured *Eschscholtzia californica* cells reacted with the elicitors from yeast extract *Colletotrichum lindemuthianum* and *Verticillum dahliae* (Figure 3). Elicitation in combination with two-phase culture additionally increased net sanguinarine production, as well as the sanguinarine concentration in the accumulation phase.

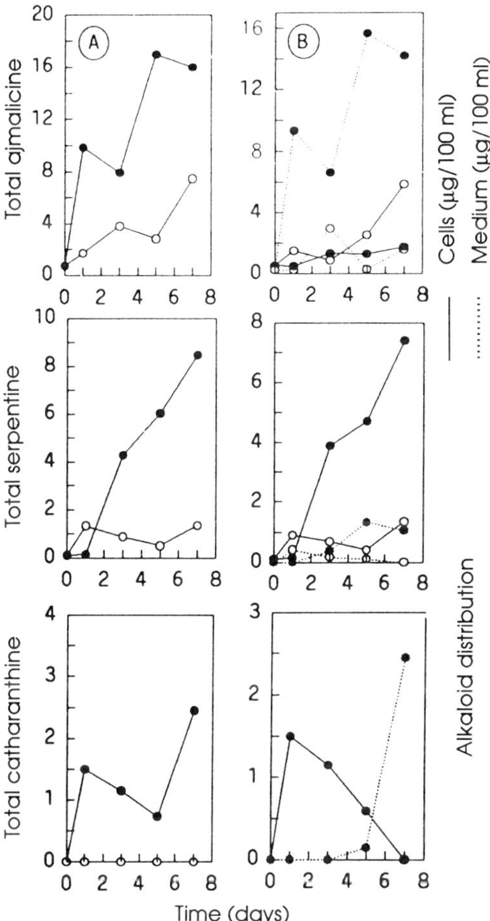

Figure 1. *Total alkaloid production (A) by* Catharanthus roseus *cells and distribution (B) between cells and medium with elicitor treatment,* ●; *without elicitor treatment,* ○. *(From Nef, C., Rio, B., and Chrestin, H.,* Plant Cell Rep., *10, 26, 1991. With permission.)*

On the other hand, elicitation with fungal elicitor changed the cell and medium distribution of alkaloids in *Catharanthus roseus* cell suspension cultures (Figure 4).[64] The distribution of the three alkaloids between cells and medium showed different long-term responses to elicitation. The different distributions of ajmalicine and catharanthine do not fit with the simple pK-dependent pH-driven diffusion transport mechanism.[65] The presence of the fungal elicitor may accelerate the physiological or the molecular processes leading to the release of catharanthine. From these studies, it is proposed that plant cell suspension cultures with fungal elicitor will be good tools for studying the molecular basis of the expression of secondary metabolites synthesis and control of its

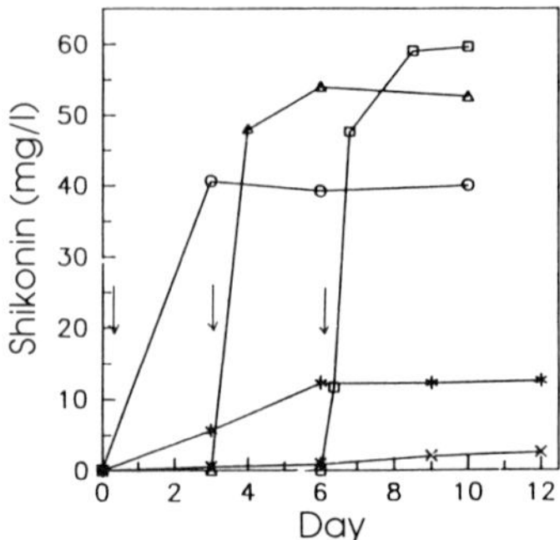

Figure 2. *Time courses of shikonin production of* Lithospermum erythrorhizon *suspension cultures (*in situ *extraction and elicitor treatment): elicitor and solvent addition time, 0 day, ○; 3 day, △; 6 day, □;* in situ *extraction, control culture,*. (From Kim, D. J. and Chang, H. N.,* Biotechnol. Lett., *12, 6, 443, 1990. With permission.)*

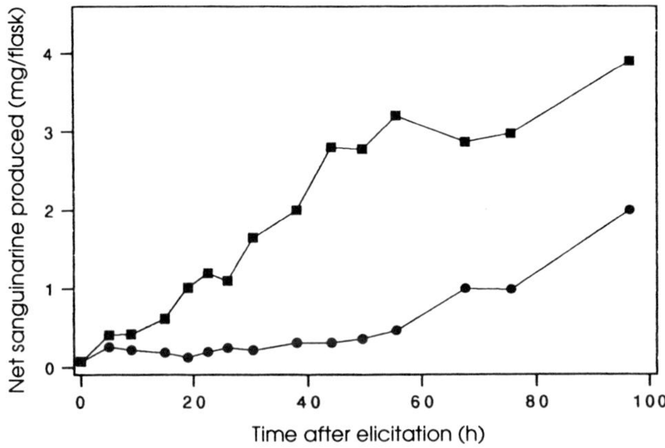

Figure 3. *Elicitation of total sanguinarine production in two-phase culture: with elicitation, ■; without elicitation, ●. (From Byun, S. Y., Ryu, Y. W., Kim, C., and Pedersen, H.,* J. Ferm. Technol., *5, 380, 1992. With permission.)*

compartment. In addition, the release of considerable proportions of the accumulated secondary metabolites suggests the potential of this approach for the generation of secondary metabolites and the study of their synthesis.

Conclusions

In order to achieve the accumulation of secondary metabolites, the use of plant cell suspension cultures with elicitor will become increasingly important

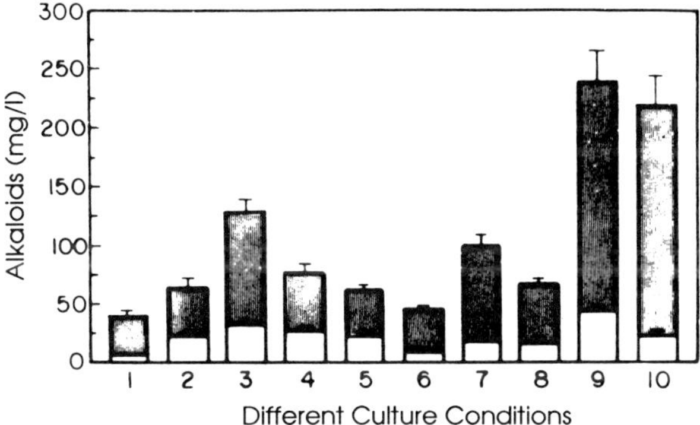

Figure 4. *Effect of the addition of* Aspergillus niger *homogenates and enzyme inhibitors and the use of osmotic stress for 72 hours on the total alkaloid production of* Catharanthus roseus *suspension culture line BAP-2: alkaloids in the cell,* □*; alkaloids released to the medium,* ■. *1 = control at day 16; 2 = fungal homogenate; 3 = transcinnamic acid; 4 = fungal homogenate + transcinnamic acid; 5 = fungal homogenate + caffeic acid; 6 = fungal homogenate + caffeic acid + tryptophan; 7 = osmotic stresses; 8 = control at day 19; 9 = fungal homogenate + transcinnamic acid for 72 hours and osmotic stresses for 72 hours; 10 = osmotic stress for 72 hours and fungal homogenate + transcinnamic acid for 72 hours. (From Godoy-Hernandez, G. and Loyola-Vargas, V. M.,* Plant Cell Rep., *10, 537, 1991. With permission.)*

in the future, because cultured plant cells are potentially valuable sources of pharmaceuticals and novel biological compounds. The ability of certain products or groups of products to be accumulated as a result of the *de novo* stimulation of their biosynthesis by abiotic and/or biotic factors has been well-established for a number of cell cultures. Among them, elicitors can induce rapid and drastic increases in yields of secondary metabolites. In addition, elicitation with *in situ* extraction is believed to eliminate feedback inhibition by the secondary metabolites produced. So, elicitation with other stresses or *in situ* extraction significantly increases secondary metabolites production. Furthermore, it is expected that elicitation, combined with other stress factors, may overcome some disadvantages associated with low yields in secondary metabolites or difficulties of product separation.

References

1. **Tabata, M. and Fujita, Y.,** *Biotechnology in Plant Science,* Zaitlin, M., Day, P., and Hollaender, A., Eds., Academic Press, New York, 1985, 217.

2. **Discosmo, F. and Misawa, M.,** Eliciting secondary metabolism in plant cell cultures, *Trends Biotechnol.,* 3, 318, 1985.

3. **Mansfield, J. W., Hargreaves, J. A., and Boyle, F. C.,** Phytoalexin production by live cells in broad bean leaves infected with *Botrytis cinerea, Nature,* 252, 316, 1974.

4. **Tomiyama, K.,** Double infection by an incompatible race of *Phytophthora infestans* of a potato plant cell which has previously been infected by a compatible race, *Ann. Phytopathol. Soc. Jpn.,* 32, 181, 1966.

5. **Tomiyama, K.,** Further observation on the time requirement for hypersensitive cell death of potatoes infected by *Phytophthora infestans* and its relation to metabolic activity, *Phytopathol. Z.,* 58, 367, 1967.

6. **Keen, N. T., Pariridge, J. E., and Zake, A. L.,** Pathogen-produced elicitor of a chemical defense mechanism in soybeans monogenetically resistant to *Phytophthora megasperma* var. *sojae, Phytopathology,* 62, 768, 1972.

7. **West, C. A.,** Fungal elicitors of the phytoalexin response in higher plants, *Naturwissenschaften,* 68, 447, 1981.

8. **Threfell, D. R. and Whitehead, I. M.,** *Manipulating Secondary Metabolism in Culture,* Robins, R. J., and Rhodes, M. J. C., Eds., Cambridge University Press, Cambridge, 1988, 51.

9. **Van Der Heijden, R., Verpoorte, R., and Harkers, P. A. A.,** *Manipulating Secondary Metabolism in Culture,* Robins, R. J., and Rhodes, M. J. C., Eds., Cambridge University Press, Cambridge, 1988, 217.

10. **Sharp, J. K., McNeil, M., and Albersheim, P.,** The primary structures of one elicitor-active and seven elicitor-inactive hexa(β-D-glucopyranosyl)-D-glycitols isolated from the mycelial walls of *Phytophthora megasperma* f. sp. *glycinea, J. Biol. Chem.,* 259, 11321, 1984.

11. **Flor, H. H.,** Current status of the gene-for-gene concept, *Annu. Rev. Phytopathol.,* 9, 275, 1971.

12. **Heath, M. C.,** A generalized concept of host-parasite specifity, *Phytopathology,* 71, 1121, 1981.

13. **Ayers, A. R., Valent, B., Ebel, J., and Albersheim, P.,** Host-pathogen interactions. XI. Composition and structure of wall-released elicitor fractions, *Plant Physiol.,* 57, 766, 1976.

14. **De Wit, P. J. G. M. and Flach, W.,** Differential accumulation of phytoalexins in tomato leaves, but not in fruits after inoculation with virulent and avirulent races of *Cladosporium fulvum, Physiol. Plant Pathol.,* 15, 257, 1979.

15. **Eilert, U., Kurz, W. G. W., and Constabel, F.,** Stimulation of sanguinarine accumulation in *Papaver somniferum* cell cultures by fungal elicitors, *J. Plant Physiol.,* 119, 65, 1985.

16. **Heinstein, P. F.,** Future approaches to the formation of secondary natural products in plant cell suspension cultures, *J. Nat. Prod.,* 48, 1, 1985.

17. **Eilert, U., Constabel, F., and Kurz, W. G. W.,** Elicitor-stimulation of monoterpene indole alkaloid in suspension cultures of *Catharanthus roseus, J. Plant Physiol.,* 126, 11, 1986.

18. **Fukui, H., Yoshikawa, N., and Tabata, M.,** Induction of shikonin formation by agar in *Lithospermum erythrorhizon* cell suspension cultures, *Phytochemistry,* 22, 11, 2451, 1983.

19. **Fukui, H. and Tani, M.,** Induction of shikonin biosynthesis by endogeneous polysaccharides in *Lithospermum erythrorhizon* cell suspension cultures, *Plant Cell Rep.,* 9, 73, 1990.

20. **Kim, D. J. and Chang, H. N.,** Increased shikonin production in *Lithospermum erythrorhizon* suspension cultures with *in situ* extraction and fungal cell treatment (elicitor), *Biotechnol. Lett.,* 12, 6, 443, 1990.

21. **Kohle, H., Young, D. H., and Kauss, H.,** Physiological changes in suspension cultured soybean cells elicited by treatment with chitosan, *Plant Sci. Lett.,* 33, 221, 1984.

22. **Hille, A., Purwin, C., and Ebel, J.,** Induction of enzymes of phytoalexin synthesis in cultured soybean cells by an elicitor from *Phytophthora megasperma* f. sp. *glycinea, Plant Cell Rep.,* 1, 123, 1982.

23. **Robbins, M. P., Bolwell, G. P., and Dixon, R. A.,** Metabolic changes in elicitor-treated bean cells. Selectivity of enzyme induction in relation to phytoalexin accumulation, *Eur. J. Biochem.,* 148, 563, 1985.

24. **Dixon, R. A. and Bendall, D. S.,** Changes in phenolic compounds associated with phaseollin production in cell suspension cultures of *Phaseolus vulgaris, Physiol. Plant Pathol.,* 13, 283, 1978.

25. **Dixon, R. A. and Fuller, K. W.,** Characterization of components from culture filtrates of *Botrytis cinerea* which stimulate phaseollin biosynthesis in *Phaseolus vulgaris* cell suspension cultures, *Physiol. Plant Pathol.,* 11, 287, 1977.

26. **Dixon, R. A., Dey, P. M., Murphy, D. L., and Whitehead, I. M.,** Dose responses for *Colletotrichum lindemuthianum* elicitor-mediated enzyme induction in French bean cell suspension cultures, *Planta,* 151, 272, 1981.

27. **Tietjen, K. G., Hunkler, D., and Matern, U.,** Differential response of cultured parsley cells to elicitors from two non-pathogenic strains of fungi. I. Identification of induced products as coumarine derivatives, *Eur. J. Biochem.,* 131, 401, 1982.

28. **Jahnen, W. and Hahlbrock, K.,** Cellular localization of nonhost reactions of parsley *(Petroselinum crispum)* to fungal infection, *Planta,* 173, 197, 1988.

29. **Eilert, U., Ehmke, A., and Wolters, B.,** Elicitor-induced accumulations of acridone alkaloid epoxides in *Ruta graveolens* suspension cultures, *Planta Med.,* 50, 507, 1984.

30. **Wolters, B. and Eilert, U.,** Accumulation of acridone-epoxides in callus cultures of *Ruta graveolens* increased by mixed culture with fungi, *Z. Naturforsch.,* 37, 575, 1982.

31. **Funk, C., Gugler, K., and Brodelius, P.,** Increased secondary product formation in plant cell suspension cultures after treatment with a yeast carbohydrate preparation (elicitor), *Phytochemistry,* 26, 2, 401, 1987.

32. **Byun, S. Y., Ryu, Y. W., Kim, C., and Pederson, H.,** Elicitation of sanguinarine production in two phase cultures of *Eschscholtzia californica, J. Ferm. Bioeng.,* 73, 5, 380, 1992.

33. **Whitehead, I. M., Atkinson, A. L., and Threlfall, D. R.,** Studies on the biosynthesis and metabolism of the phytoalexin lubimin and related compounds in *Datura stramonium* L., *Planta,* 182, 81, 1990.

34. **Sitton, D. and West, C. A.,** Casbene, an antifungal diterpene produced in cell-free extracts of *Ricinus communis* seedlings, *Phytochemistry,* 14, 1921, 1975.

35. **Kessmann, H., Choudhary, A. D., and Dixon, R. A.,** Stress responses in alfalfa. III. Induction of medicarpin and cytochrome P450 enzyme activities in elicitor-treated cell suspension cultures and protoplasts, *Plant Cell Rep.,* 9, 38, 1990.

36. **Sharp, J. K., Valent, B., and Albersheim, P.,** Purification and partial characterization of a β-glucan fragment that elicits phytoalexin accumulation in soybean, *J. Biol. Chem.,* 259, 11312, 1984.

37. **Sequeira, L.,** Mechanisms of induced resistance in plants, *Annu. Rev. Microbiol.,* 37, 51, 1983.

38. **Boller, T.,** Hydrolytic enzymes in plant disease resistance, in *Plant-Microbe Interaction: Molecular and Genetic Perspectives 2,* Kosuge, T. and Nester, E. W., Eds., Macmillan, New York, 1987, 385.

39. **Paxton, J. D.,** *Phytophthora* root and stem rot of soybean: a case study, in *Biochemical Plant Pathology,* Callow, J. A., Ed., John Wiley & Sons, Chichester, UK, 1983, 19.

40. **Jahnen, W. and Hahlbrock, K.,** Cellular localization of non-host resistance reactions of parsley *(Petroselinum crispum)* to fungal infection, *Planta,* 173, 197, 1988.

41. **Hahn, M. G., Bonhoff, A., and Grisebach, H.,** Quantitative localization of the hytoalexin glyceollin I in relation to fungal hyphae in soybean roots infected with *Phytophthora megasperma* f. sp. *glycinea, Plant Physiol.,* 77, 591, 1985.

42. **Scheel, D., Hauffe, K. D., Jahnen, W., and Hahlbrock, K.,** Stimulation of phytoalexin formation in fungus-infected plants and elicitor-treated cell cultures of parsley, in *Recognition in Microbe-Plant Symbiotic and Pathogenic Interactions,* Lugtenberg, B., Ed., Springer, Berlin, 1986, 325.

43. **Park, J. E., Hahlbrock, K., and Scheel, D.,** Different cell-wall components from *Phytophthora megasperma* f. sp. *glycinea* elicit phytoalexin production in soybean and parsley, *Planta,* 176, 75, 1988.

44. **Schmidt, W. E. and Ebel, J.,** Specific binding of a fungal glucan phytoalexin elicitor to membrane fractions from soybean *Glycine max, Proc. Natl. Acad. Sci. U.S.A.,* 84, 4117, 1987.

45. **Kauss, H., and Jeblick, W.,** Solubilization, affinity chromatography and Ca^{2+}/polyamine activation of the plasma membrane activated 1,3-β-D-glucan synthase, *Plant Sci.,* 48, 63, 1987.

46. **Atkinson, M. M., Huang, J. S., and Knopp, J. A.,** The hypersensitive reaction of tobacco to *Pseudomonas syringae* pv. *pisi* activation of a plasmalemma K^+/H^+ exchange mechanism, *Plant Physiol.,* 79, 843, 1981.

47. **Pelissier, B., Thiband, J. B., Grignon, C., and Esquerre-Tugye, M. T.,** Cell surfaces in plant-microorganism interactions. VII. Elicitor preparations from two fungal pathogens depolarize plant membranes, *Plant Sci.,* 46, 103, 1986.

48. **Farmer, E. E., Pearce, G., and Ryan, C. A.,** *In vitro* phosphorylation of plant plasma membrane proteins in response to the proteinase inhibitor inducing factor, *Proc. Natl. Acad. Sci. U.S.A.,* 86, 1539, 1989.

49. **Dietrich, A., Mayer, J. E., and Hahlbrock, K.,** Fungal elicitor triggers rapid, transient, and specific protein phosphorylation in parsley cell suspension cultures, *J. Biol. Chem.,* 265, 6360, 1990.

50. **Felix, G., Grosskopf, D. G., Regenass, M., and Boller, T.,** Rapid changes of protein phosphorylation are involved in transduction of the elicitor signal in plant cells, *Proc. Natl. Acad. Sci. U.S.A.,* 8831, 1991.

51. **Hahlbrock, K. and Grisebach, H.,** Enzyme controls in the biosynthesis of lignins and flavonoids, *Annu. Rev. Plant Physiol.,* 30, 105, 1979.

52. **Hahlbrock, K.,** Flavonoids, in *The Biochemistry of Plants,* 7, Stumpf, P. K. and Conn, E. E., Eds., Academic Press, New York, 1981, 425.

53. **Nover, L., Ed.,** *Heat Shock Response of Eucaryotic Cells,* Springer-Verlag, Berlin, 1984.

54. **Kimpel, J. A. and Key, J. L.,** Heat shock in plants, *Trends Biochem. Sci. U.S.A.,* 10, 353, 1985.

55. **Ashburner, M. and Bonner, J. J.,** The induction of gene activity in *Drosophila* by heat-shock, *Cell,* 17, 241, 1979.

56. **Findly, R. C. and Pederson, T.,** Regulated transcription of the genes for actin and heat-shock proteins in cultured *Drosphila* cells, *J. Cell Biol.,* 88, 323, 1981.

57. **Walter, M. H.,** The induction of phenylpropanoid biosynthetic enzymes by ultraviolet light or fungal elicitor in cultured parsley cells is overriden by a heat-shock treatment, *Planta,* 177, 1, 1989.

58. **Furuya, T., Ikuta, A., and Syono, K.,** Alkaloids from callus tissue of *Papaver somniferum, Phytochemistry,* 11, 3041, 1972.

59. **Cline, S. D. and Coscia, C. J.,** Stimulation of sanguinarine production by combined fungal elicitation and hormonal deprivation in cell suspension cultures of *Papaver bracteatum, Plant Physiol.,* 86, 160, 1988.

60. **DiCosmo, F. and Towers, G. H. N.,** Stress and secondary metabolism in cultured plant cells, in *Recent Advances in Phytochemistry,* 18, Timmermann, B. N., Steelink, C., and Loewus, F. A., Eds., Plenum Press, New York, 1984, 97.

61. **Godoy-Hernandez, G. and Loyola-Vargas, V. M.,** Effect of fungal homogenate, enzyme inhibitors and osmotic stress on alkaloid content of *Catharanthus roseus* cell suspension cultures, *Plant Cell. Rep.,* 10, 537, 1991.

62. **Felix, H., Brodelius, P., and Mosbach, K.,** Enzyme activities of the primary and secondary metabolism of simultaneously permeabilized and immobilized plant cells, *Anal. Biochem.,* 116, 462, 1981.

63. **Berlin, J., Witte, L., Schubert, W., and Wray, V.,** Determination and quantification of monoterpenoids secreted into the medium of cell cultures of *Thuja occidentalis, Phytochemistry,* 23, 1277, 1984.

64. **Nef, C., Rio, B., and Chrestin, H.,** Induction of catharanthine synthesis and stimulation of major indole alkaloids production by *Catharanthus roseus* cells under non-growth altering treatment with *Pythium vexans* extracts, *Plant Cell Rep.,* 10, 26, 1991.

65. **Bouyssou, H., Pareilleux, A., and Marigo, G.,** The role of pH gradients across the plasmalemma of *Catharanthus roseus* and its involvement in the release of alkaloids, *Plant Cell Tissue Organ Cult.,* 10, 91, 1987.

Transformation of Plant Cells by Bombardment with Microprojectiles

R. N. Chibbar and K. K. Kartha

Plant Biotechnology Institute, National Research Council of Canada, Saskatoon, Saskatchewan, Canada

Introduction

Transformation is defined as a permanent genetic change in a cell following incorporation of foreign DNA. The principle of introducing alien genes, conferring useful traits, into a crop plant has been successfully exploited by plant breeders to improve agricultural crops. Several novel varieties of crops with improved agronomic performance have been produced by conventional plant breeding. Despite the past success and the variety of techniques currently available to plant breeders, they are limited by several problems.[1] One of the major problems facing plant breeders is the dwindling gene pool that is available for manipulation through sexual crosses.[2] It is in this context that transformation techniques to introduce alien genes nonsexually into plants have gained importance.

0-8493-8262-9/94/$0.00+$.50

Particle Bombardment

The Process

A variety of techniques is available for introducing DNA into plants; however, the most recent technique and the topic of this chapter is *microprojectile bombardment.* This technique has been termed *biolistics* by the inventors[3] and is also known as particle bombardment, particle acceleration, the gene gun, or the particle gun method. The process, as described by the inventors, essentially consists of accelerating particles to a speed at which they can penetrate the surface of the cell and be incorporated into the interior of the cell.[4] When applied to the genetic transformation of plant cells, inert gold or tungsten particles are coated with genetic material capable of functioning in plant cells and are delivered into plant cells, where the genetic material is incorporated into the cell's nuclear genome and functions as any other existing plant gene.

Apparatus

The particle bombardment apparatus essentially consists of a mechanism to accelerate the particles to the desired velocities and to regulate their penetration into the recipient cells. A few designs have been reported in the literature that vary in these two major aspects.[5–14] The original apparatus designed by the researchers at Cornell utilized a gunpowder discharge to accelerate microprojectiles coated with biologically active material. The basic principle, as well as the description of the equipment, has been detailed by the inventors.[5,6] A brief description of the currently used helium-driven gun[8] (Biolistics® PDS-1000/He) is given below. The main component of this system is a rupture disc assembly that controls the helium pressure at which it propels the microprojectiles that carry the vector DNA (Figure 1). The rupture disc assembly consists of a gas acceleration tube with a rupture disc inside a retaining cap on the bottom of the tube. The microprojectiles coated with DNA are placed on a carrier which fits below the rupture disc. The chamber is partially evacuated and the helium pressure is allowed to build up until the rupture disc ruptures. The microprojectiles are then propelled by the helium pressure which delivers them through a metal screen to hit the target cells.

An electric arc discharge gun has also been reported by Christou et al.[9] This gun uses shock waves generated by an electrical discharge to propel the gold particles/DNA suspension towards the target cells. The propulsion mechanism consists of a polyvinyl chloride pipe (13 mm inner diameter or i.d.) that has 2 electrodes 5 mm below its end. Power is provided to the electrodes by a capacitor for which the charge can be regulated by a DC power supply. When a current is applied, the spark gap is broken by a droplet of water, creating sufficient energy to propel the particles at high velocities. Aluminum foil carrying the gold particles and DNA is supported on an expansion spacer

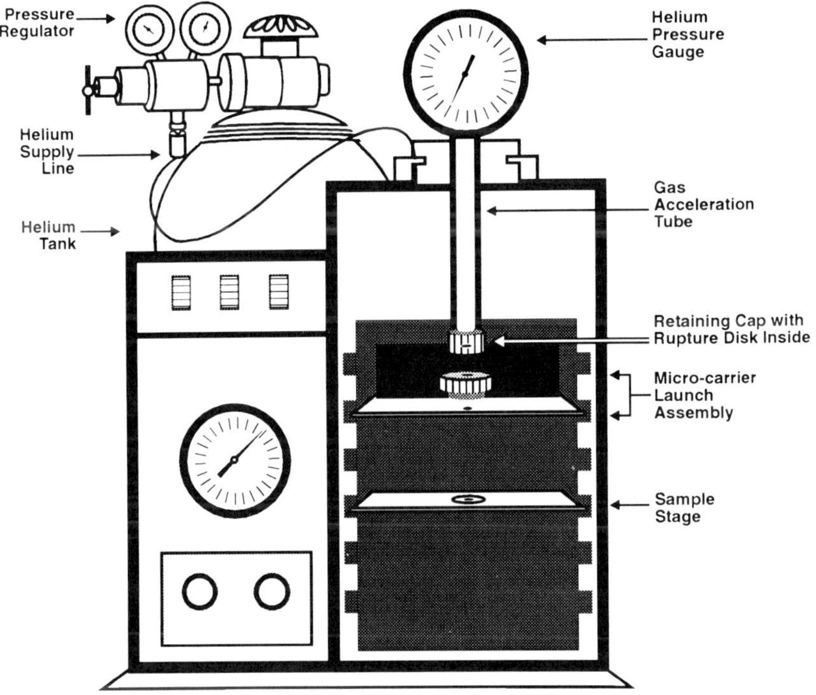

Figure 1. *A diagrammatic representation (not to scale) of a helium-powered Biolistics® particle gun. (Reproduced with permission. The PDS-1000/He is manufactured by Biorad, Life Sciences Division, Hercules, California.)*

placed above the electrodes. A 100 µm mesh screen suspended above the gun catches the carrier sheet but allows the accelerated gold particles to pass through and impact the target tissue.

Uneven particle penetration of the target cells or tissues has been a major drawback of the particle bombardment apparatuses. Attempts to regulate particle penetration have focused on modifications of the power source used to propel the microprojectiles carrying the vector DNA. Regulated nitrogen gas pressure[10] or compressed air-driven[11] equipment has been reported. Oard et al.[12] used an air gun to propel the microprojectiles. Finer et al.[13] have developed a *particle in flow gun* (PIG) in which microprojectiles are accelerated directly into a stream of helium rather than being supported by a macrocarrier.

Sautter et al.[14] have used an air gun to generate a pressure pulse of approximately 2 msec at a pressure of 60 bar to deliver the microprojectile/DNA suspension. However, the main difference between this and other systems is that the DNA is suspended with gold microprojectiles rather than being coated on them. The droplets containing these two components impact the target cells, and the gold particles create holes through which the DNA passes into the cells.

According to the inventors, the fact that DNA moves independently of the particles is a major advantage, in that they can microtarget small areas of tissues.

Factors Influencing Gene Delivery

The main objective of microprojectile bombardment is to deliver an optimal amount of DNA into plant cells with minimal cell damage. Since both the terms 'optimum' and 'damage' are subjective and vary from cell to cell, it is not possible to arrive at ideal conditions that are applicable to all plant cells. The various factors influencing the delivery of genes into maize cells using this technology have been discussed by Klein et al.[15] and in subsequent reviews.[16-18] Recently, Sanford et al.[19] described in detail the various factors and experiments used to optimize the gene delivery into different plant cell and tissue recipient systems. Factors affecting the delivery of foreign DNA into plant cells are optimized by evaluating the expression of the introduced genes. The important factors that influence the delivery of foreign DNA into plant cells are discussed in the following sections.

Microprojectiles — The main criteria for the suitability of a microprojectile are that it be an inert material of high density and that it be capable of retaining precipitated macromolecules — in this case, DNA. Tungsten and gold have been the two elements of choice, since they meet these basic criterion. Tungsten is relatively less expensive as compared to gold and the microprojectiles can be obtained in various sizes (0.7 to 4.0 μm); however, a high level of cell death due to tungsten toxicity has been reported.[20] This was inferred from the tungsten-induced inhibition of plant cell growth *in vitro*. A possible cause is the lowering of pH by tungsten. However, buffering of the medium did not alleviate the tungsten-induced inhibition of growth. The growth inhibitory effect of tungsten is also dependent upon the type of organism and cells used to receive DNA.[20] Nevertheless, an overwhelming majority of reports published between 1989 and 1992 used tungsten microprojectiles in their experimentation. Similarly, in our experience with wheat and barley cell cultures and immature zygotic embryos, we did not notice any obvious growth inhibitory effects of tungsten.[21–23] In the recent reports, gold particles are used primarily because of their nontoxic nature in plant cells. Gold particles are more homogenous in size (1 to 5 μm) and shape (circular) as compared to tungsten particles. The only disadvantage is the cost of gold particles and limitations imposed by their size (>1 μm) for specialized applications. Besides these two most commonly employed microprojectiles (tungsten and gold), the use of dried *Escherichia coli* bearing plant vectors and glass fragments has also been suggested.[19]

Attachment of vector DNA to microprojectiles — DNA is commonly attached to the microprojectiles using calcium chloride and spermidine precipitation.[24] In a recent study, Perl et al.[25] bombarded wheat callus in the presence of silver thiosulfate and calcium nitrate, rather than calcium chloride, and eliminated spermidine from the DNA attachment mixture. This procedure gave severalfold higher transient expression of β-glucuronidase *(gus)* reporter gene in wheat scutellar callus, as compared to the conventional calcium chloride and spermidine attachment protocol of DNA to microprojectiles. A higher pH (10) also increased the transient expression of *gus* gene in the same tissue,[25] which stresses the importance of optimizing various parameters to obtain maximal gene expression.

Biolistic parameters — The biolistics parameters affect the delivery of DNA into the plant cells and the injury it causes to those cells during that process. Using an optimized protocol, the maximum amount of DNA is delivered with minimal injury to the recipient cells. The pressure with which the microprojectiles are propelled toward the target cells and the distance that the particles travel before striking the target (flight distance) both affect the extent of cell injury. For example, with cultured barley cells bombarded using a helium-driven particle gun, maximum transient expression of *gus* gene was observed with a pressure of 1300 or 1800 psi and a particle flight distance of 9.5 cm (Figures 2 and 3). However, cells bombarded with higher pressure or a shorter flight distance showed no GUS activity in the center of the target area, possibly due to cell death. Lower pressure or longer flight distance resulted in a reduced number of GUS expressing foci in the target cells (unpublished results and Reference 21).

Recipient systems — Any plant cell can serve as a recipient for DNA delivered via biolistics. This cell can be present as a single cell or can be within a cluster of cells in suspension culture or callus or in an organized tissue. Expression of marker genes introduced via biolistics has been shown in intact meristems,[26–28] leaves,[29–31] roots,[31] hypocotyl sections,[32] cotyledons,[33] zygotic embryos,[21,22,34] callus,[25,35] suspension culture cells,[21,24,35–37] and somatic embryos.[38,39] However, transient expression has always been higher in suspension-cultured cells as compared to organized tissues.[21] Similarly, the frequency of obtaining stable transformed cells has also been higher in suspension-cultured cells as compared to other organized tissues, such as immature zygotic embryos. Yamashita et al.[40] observed only one gold microprojectile in each of the majority of the cultured tobacco cells that had been bombarded with GUS expression cassettes coated on gold microprojectiles. The majority (70%) of the microprojectiles were present in the vacuole, and a mere 7 to 8% were imbedded in the nucleus. However, 90% of the cells that had the microprojectiles in the nucleus showed GUS activity. Therefore, they suggested that cells undergoing mitotic divisions are ideal recipients of microprojectile-delivered DNA because of the absence of nuclear membrane at this stage. Alternatively, the vacuole may be reduced by treating the cells with a high osmoticum. This may also explain the higher level of transient gene expression observed in actively dividing cells in suspension culture as compared to organized tissues.[21]

Injury to the recipient system — The main area of concern with biolistics is the injury to the recipient system. Taylor and Vasil[41] have defined two types of injury to the recipient system: pit damage and shock injury. Pit damage is

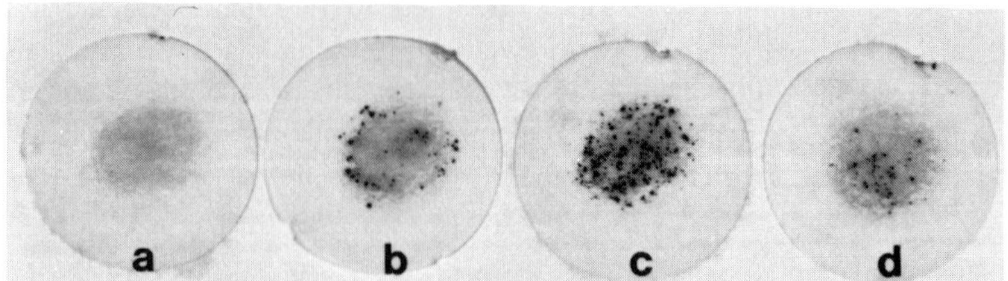

Figure 2. *Effect of flight distance of microprojectiles on transient expression of* gus *gene in cultured wheat cells: (a) control; (b) 6.5 cm; (c) 9.5 cm; (d) 13 cm. Cells on filter papers were incubated, 48 hours after bombardment, with X-gluc stain for 24 hours.*[22]

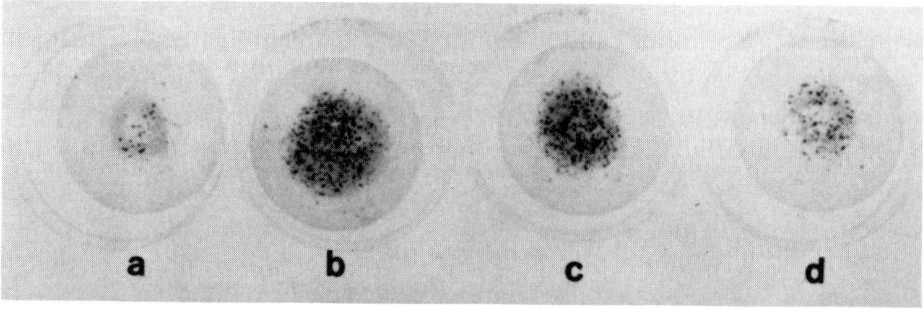

Figure 3. *Effect of helium pressure (psi) on transient expression of* gus *gene in cultured wheat cells: (a) = 450; (b) = 1300; (c) = 1800; (d) = 2200 psi. The time and method of determination of GUS activity are the same as in Figure 2.*

induced by the aggregation of the microprojectiles — in this case, gold particles — at the bottom of the pits. Shock damage is induced by the preponderance of lipids in the cells obscuring the other organelles. Shock damage was observed even in the absence of microprojectiles, suggesting that the shock force of bombardment was responsible for this effect. The damage to the cells was greater when the microprojectile flight distance was short. Similar results have been observed by Russel et al.[20] The central epicenter of the blast showed maximal cell death, and they called this area a *zone of death*. The use of post-launch baffles or meshes has been suggested to moderate the acoustic shock, to break the microprojectile aggregates, and to focus the zone of transformation.[20] Most of these problems have been overcome by the helium-powered device. The chance of recovering stable transformants in this central zone of death is less as compared to cells peripheral to this zone. However, in our experiments using cultured wheat and barley cells, we have recovered stable transformants from cells distributed all over the plate.[23] Perl et al.[25] reported a reduction in both types of damage by reducing the formation of aggregates of the microprojectiles.

Genetic Transformation of Plants

The ultimate objective of any gene transfer method is to produce stable transformed cells and plants. Stably transformed cell lines using biolistics have been demonstrated in various plants such as tobacco,[42] corn,[43] wheat,[23,44] and sorghum.[45] Furthermore, particle bombardment has been successfully used to produce transgenic cereals such as corn,[46–48] rice,[49–51] and wheat.[52–54] However, production of transgenic plants is a result of successful interplay among several factors, some of which are discussed in the following sections.

Target Tissues for Production of Transgenic Plants

Various types of plant cells and organized tissues have been used to produce transgenic plants by means of biolistics for the gene insertion. Cultured cells are the most commonly used cell recipient system for the successful demonstration of transient expression of genes introduced via biolistics, as well for the production of transgenic monocot crops. As discussed earlier, cells in an active state of growth are more amenable to gene insertion and integration as compared to those in either slow-growing or quiescent stages. Embryogenic cell cultures have also been used to obtain transgenic plants via biolistics.[46–48,50,51,55–57] Suspension-cultured cells exhibited higher frequencies of gene insertion and integration as compared to embryogenic calli.[23,54] However, derivation of embryogenic cell suspension cultures is very genotype specific and extremely time-consuming; plants regenerated from cell suspension cultures are mostly sterile and abnormal; and the embryogenic potential of such cultures is lost over time upon subculture.[46–48] Embryogenic callus was employed to produce transgenic wheat[53] and sugarcane.[58] Transgenic rice has been produced by microprojectile bombardment of intact immature zygotic embryos.[49] Isolated scutella from immature zygotic embryos of wheat and barley (enhanced regeneration system or ERS) when cultured under ideal conditions become very prolific in cell division, giving rise to a large number of somatic embryos in a few weeks.[54] The frequency of gene insertion, integration, and production of transgenic wheat plants is higher in this system than in presently available embryogenic wheat callus systems.[53,54] Other tissues employed for the production of transgenic plants include apical meristems (soybeans[27,28] and common beans[26]), leaves,[29,30,59] petioles,[30] roots,[60] immature zygotic embryos,[61] somatic embryos,[38,39] and protocorms.[62]

Selection of Transgenic Cells and Tissues

The development of a suitable selection strategy is very important in any transformation protocol, especially in biolistics-mediated transformation since the stable transformation frequency is relatively low as compared to other methods of direct gene transfer. An ideal selective agent should provide a tight selection of transformed cells and tissues without compromising on regeneration and fertility of transformed plants. Kanamycin (an antibiotic of the aminoglycoside group), resistance to which is conferred by the neomycin phosphotransferase *(nptII)* gene, has been extensively used to select transformed cells and tissues in dicots. However, monocot species have a high level of inherent resistance to kanamycin,[63,64] thus limiting its use as a selection agent in these species. Consequently, the use of hygromycin, another member of the aminoglycoside group, has been proposed for selection of transformed monocot cells and tissues.[63,64] Although hygromycin has been used to select transformants in maize[48] and rice,[49] in our experiments with transformation of wheat and barley it did not prove to be satisfactory due to its interference with the morphogenetic potential of selected cells. We have found that geneticin (G418), another antibiotic of the aminoglycoside group, fulfills most of the requirements of a selective agent in wheat.[23,54] It has also been used to select transformed cells of sugarcane.[58] Resistance to L-phosphinothricin (L-PPT), conferred by phosphinothricin acetyl transferase *(bar)* gene, has been successfully used to select transformants in corn, wheat, and some other plants. The selection conferred by *bar* was not absolute since growth of many nontransformed cells occurred, especially in wheat.[53] Vasil et al.[44] have also used glyphosate resistance induced by expression of a mutant gene encoding the enzyme 5-enol-pyruvylshikimicacid-3-phosphate (EPSP) synthase to select transformed wheat cells and calli.

Plant Gene Expression Vectors

Vector DNA capable of functioning in recipient cells is a vital component for the transformation of plant cells. An important consideration is the choice of a promoter that directs the expression of introduced genes. Promoters from different organisms such as bacteria, viruses, and plants have been used to express foreign genes in plants. The commonly used promoters have been derived from the T-DNA genes for opine synthesis,[65] nopaline synthesis,[66] and divergent dual promoters of mannopine synthase genes 1 and 2[67] and the 35S RNA promoter of the cauliflower mosaic virus (35S).[68] The 35S promoter has been most commonly employed to obtain transgenic plants. The promoter fragment is 343 bp long and contains a strong transcriptional enhancer.[69–71] Duplication of this enhancer fragment (E35S) gives tenfold higher gene expression in dicot cells such as tobacco.[72] The activity of 35S promoter has been considered to be constitutive as a result of the interaction of different domains in the promoter. These domains independently direct tissue specific expression.[73] Although the 35S promoter has been successfully used in both dicots and monocots to produce transgenic plants, it is considered as a weak promoter for monocots. In a recent report, four different promoter sequences were compared to drive the expression of *gus* gene in cultured barley cells, and it was found that 35S promoter was the least effective in directing *gus* expression.[74]

Studies have been carried out to identify other promoters to drive foreign gene expression in monocot cells. Maize alcohol dehydrogenase 1 *(Adh1)* promoter[75] has been used to produce transgenic rice and wheat callus lines. The maize *Adh1* promoter is anaerobically induced, and genes under its control are not always expressed. The anaerobic response element[76,77] (ARE) identified from this promoter has been used to construct a synthetic promoter pEMU[78] that directs a higher level of gene expression in monocot cells. The pEMU promoter is based on a truncated maize *Adh1* promoter with six ARE sequences and four octopine synthase[79] (OCS) enhancer elements. This promoter, in combination with *Adh1* intron 1, gave 10- to 50-fold higher *gus* marker gene expression in transient assays in protoplasts of several monocot species.[78] A rice actin promoter[80] has also proven to be very efficient in directing foreign gene expression in monocot cells. The rice actin promoter-intron constructs have been improved by incorporating an optimal translation context for *gus* gene. These constructs give severalfold higher expressions of *gus* in various monocot cells as compared to other promoters tested.[81] The 5′ untranslated region of maize ubiquitin gene, along with its first intron when fused with *cat* gene, exhibited tenfold higher CAT activity in maize protoplasts than the constructions with the 35S promoter.[82] Similarly, the promoter region from a rice *GOS2* gene has been shown to be more active as compared to the 35S promoter in barley and maize.[83]

These studies show the current emphasis on the isolation and characterization of promoters that direct the constitutive expression of introduced genes. However, with the advances in transformation technology, inducible and tissue specific promoters will be more desirable. The production of a specific transgenic plant will demand the targeting of introduced genes to specific tissues and developmental stages. A few promoters in this class have been identified and tested in transgenic dicot plants where they have maintained their specificity.[73] The biolistics technology will be very useful in assessing the functionality of the tissue, and developmental stage specific promoters as the gene constructs can be directly introduced into organized tissues and their activity monitored by transient assays in 24 to 48 h.[31]

Enhanced expression of introduced genes is desirable for various reasons. In the initial stages, a high level of expression of selectable marker genes is helpful for the selection of transformed cells under stringent selection conditions, and in the case of genes encoding agronomic traits, it is needed for the product to be formed in adequate quantities. This has been achieved by introducing leader sequences between the promoter and the coding sequence of the genes. In the case of dicots, untranslated leader sequences from viruses have proven to be most successful. For example, the 67 bp leader sequence omega (Ω) derived from tobacco mosaic virus (TMV) stimulates gene expression in animals, plants, and bacteria.[84] Similarly, the incorporation of the untranslated leader sequence from RNA4 of alfalfa mosaic virus (AMV) gave a 35-fold increase in mRNA translation efficiency.[85] The inclusion of this AMV leader sequence between the *nos,* 35S, and E35S promoters, respectively, significantly enhanced the transient expression of introduced *gus* gene in various pea tissues.[31] The 5′ untranslated regions from other viruses (such as pea seed borne mosaic poty virus and tobacco mosaic virus) also enhanced the expression of foreign genes in tobacco and pea protoplasts.[86] Inclusion of a dicot intron with the RNA leader sequence Ω, derived from tobacco mosaic virus between the promoter and the coding region of the marker gene, enhanced gene expression by tenfold in electroplated bean protoplasts.[87]

An intron positioned between the promoter and the 5′ end of the marker gene significantly enhances the expression of introduced genes in monocot cells as determined by transient assays.[88] The commonly used introns that enhance the expression of introduced genes are: maize *Adh1* introns 1, 2, and 6;[88,89] maize shrunken locus 1 *(sh1)* intron 1;[90,91] the maize intron from Bronze locus;[88] first intron from rice actin 1[80,81] and actin 3;[92] and the intron from maize 82 kD heat shock protein.[93] The first untranslated 52 bp long exon of *sh1* locus also enhances the foreign gene expression, and the effect is multiplicative in combination with *sh1* first intron.[91] Although the precise mechanism of enhancement of gene expression by introns is not clear, in all studies an increased level of mRNA was observed.[88,91,94] These studies demonstrate the various approaches that are available to enhance foreign gene expression in monocots.

Integration and Inheritance of Introduced Genes

Various criteria are employed to confirm transformation. However, the first proof of stable transformation is the presence of the introduced DNA in the host genome. This is proven using a polymerase chain reaction, dot blot, or Southern analysis. The Southern analysis — employing a judicious choice of restriction enzymes to digest the DNA — provides better evidence in terms of the physical state of DNA, copy number, and number of integration events. The integration patterns of the introduced genes vary in each individual transformant that is generated by biolistics. The number of copies of introduced genes that have been inserted at one or more than one site in the host genome ranges from as few as 1 to more than 20. Marker genes such as *nptII, bar, als, aphII,* and *gus* have been shown to be functional in monocots, and their physical presence and integration into the host genome has been demonstrated by Southern analysis. However, a direct correlation between the copy number and the activity conferred by the gene has not been clearly observed. This lack of direct correlation may be due to the position effect induced by the site of insertion in the host genome. The number of insertions into the host genome, as well as the activity of the introduced gene, is much higher in cultured cells as compared to organized tissues. The integration pattern of the introduced genes remains the same after successive subcultures of callus and in the progeny of plants, thus inferring that no genomic rearrangements have occurred after the primary transformation events.

The final proof of transformation is obtained by analyzing the progeny of primary transformants and demonstrating the transmission of the introduced gene to the progeny in a Mendelian fashion. Progeny analysis has been carried out in rice,[49] maize,[46–48] and wheat.[53,54] Transgenic maize[46,47] and wheat[53] plants produced by the particle bombardment of suspension cultures were male sterile and had to be pollinated by wild type. However, as suggested by these authors, the sterility was induced by the prolonged culture of cells, rather than the particle bombardment process. From our laboratory, self-fertile transgenic wheat plants have been reported[54] using an ERS in which a short *in vitro* culture period was employed. The transgenic plants obtained from this system, using two selectable markers, were phenotypically normal and transmitted the introduced genes into the progeny in a Mendelian fashion. Similarly, in rice, phenotypically normal and fertile transgenic plants were obtained by bombarding immature zygotic embryos.[49] A detailed progeny (R_1 and R_2) analysis carried out in four transgenic maize lines demonstrated the inheritance of a functional *bar* gene in a Mendelian fashion.[96] The functional *bar* gene segregated as a single unit in the four lines that were analyzed. The nonfunctional *bar* or *gus* genes were also inherited by all four lines and these appeared to be linked with the functional *bar* units. The inability of *gus* gene to express in monocots has been reported and has been ascribed to the methylation of *gus*[42] or to promoter activity.[53] On the other hand, a high level of GUS activity was observed in various parts of a wheat plant transformed with a *gus:npt* fusion gene driven by rice actin promoter.[54]

Gene Expression and Regulation: Application of Particle Bombardment

One distinct advantage of biolistics-mediated DNA delivery is its ability to deliver DNA into intact tissues and cells. The transient expression of the introduced genes can be assayed 24 to 48 h after bombardment, thus providing a quick means to determine the activity of gene constructs in the recipient cells. Until the advent of biolistics, transient assays for activity of foreign genes and promoters were possible only in protoplasts in which genes could be introduced chemically or by electroporation. However, it has been reported that foreign genes are not properly regulated in protoplasts.[97,98]

The anthocyanin biosynthetic genes and their 5′ regulatory regions have been used to study the activity and regulation of the promoter by particle bombardment-mediated delivery of the gene constructs directly into intact tissues.[99–101] Pigmentation in maize due to accumulation of anthocyanin is controlled by at least ten genes that encode regulatory or structural proteins of the anthocyanin biosynthetic pathway.[102] Mutant aleurone cells lacking anthocyanin (due to mutations in *A1* or *Bz1* genes) were bombarded with clones of two wild-type anthocyanin biosynthetic genes (*A1* and *Bz*). Following bombardment with these two genes, the aleurone cells developed pigmentation indicating that the mutation had been reversed.[99] To study the expression of these genes in different genetic backgrounds, gene constructions were made by fusing the luciferase marker gene between the 5′ and 3′ region of these anthocyanin biosynthetic genes. When these gene constructs were introduced into aleurone layers of permissive backgrounds, 30- to 200-fold higher expression of luciferase activity was observed. This infers that the genes delivered into the intact tissues by microprojectiles are regulated in a manner similar to the endogenous genes.[99] Goff et al.[100] have used microprojectile bombardment to study the transactivation of the structural genes by the regulatory genes (*C1*, *B*, and *R*) in the anthocyanin biosynthetic pathway. These genes encode tissue-specific regulatory proteins that have similarities to transcriptional activators. Intact maize tissues (aleurone and embryos) were sequentially or simultaneously bombarded with constructions comprised of B family genes and luciferase, a marker gene. The results suggested that the tissue specific expression induced by the various alleles of B and R genes is a result of the differential expression of functionally similar proteins. Another interesting observation reported by these researchers[100] is the activation of the Bronze or A1 promoters of the anthocyanin pathway in the bombarded embryogenic callus. They suggest using the embryogenic callus with a suitable genetic background as a preferred experimental system for studying the regulation of anthocyanin biosynthetic pathway, thus avoiding the time-consuming process of isolation of embryos and aleurone layers.

Ludwig et al.[101] have constructed constitutively expressed vectors with an entire transcript unit for one of the genes *(Lc)* of the R family of the anthocyanin biosynthetic pathway. The chimeric gene construct was introduced into various tissues of maize by microprojectile bombardment. Cell autonomous pigmentation was observed in all tissues that are not normally pigmented by this gene. They proposed the use of this gene as a scorable marker to study gene expression in maize and other monocot cells.

Particle bombardment and subsequent transient assays for marker genes have been used to study the DNA sequences involved in the phytochrome regulated expression of genes.[103] The 5′ sequences, along with a part of the structural region of the oat phytochrome gene, were fused to the chloramphenicol acetyl transferase *(cat)* gene. The gene constructs were introduced into dark-grown monocot and dicot seedlings using microprojectile bombardment. By using transient assays for CAT activity, it was shown that the introduced *cat* gene is expressed and was downregulated by white light in barley, rice, and oats, but no expression was observed in three dicots (tobacco, cucumber, and *Arabidopsis*) tested. It was also shown that the oat-phytochrome was regulated in rice in a similar manner to rice phytochrome genes.

Knudsen and Muller[104] biolistically delivered a gene construct in which the 552 bp upstream region of B1-hordein gene was used to drive the *gus* marker gene into developing barley endosperm. By employing transient assays for GUS activity they showed that this promoter was active only in sub-aleurone and starchy endosperm cells but not in cells devoid of starch. On the other hand, the 35S promoter was active in all the cells. Stimulation of activity of a high pI α-amylase promoter by gibberellic acid in rice aleurone cells was demonstrated by introducing the gene constructs into these cells by biolistics.[105] Biolistics has also been used to show the activity of the 5′ upstream region of a *GOS2* gene fused to a *gus* reporter gene in rice and other monocot cells.[83] These recent studies demonstrate the utility of biolistics in the functional characterization of various promoters in a stage- and development-specific manner.

Problems and Prospects

Introduction of DNA into intact cells by particle bombardment is a random process. Thus, inconsistencies that are observed in the efficiency with which DNA is introduced into the cells is a major limitation of the technique. The first contributing factor for the observed fluctuations in the results is the lack of precise control of the velocity with which the microprojectiles impact the target cells. Secondly, the variation in the size and shape of the microprojectiles and the amount of DNA attached to them also contributes to the observed inconsistency in the results between different experiments. In quantitative studies aimed at optimization of protocol for biolistics-mediated DNA introduction into the cells and studies on gene expression and regulation, proper controls and replications are needed.[19] These can at times be time consuming and frustrating. Nevertheless, for the demonstration of production of transgenic plants, particle bombardment-mediated DNA delivery has been very successful. For this technology to be used to develop plants carrying useful traits, a large number of transformants will have to be produced on a regular basis. To achieve this objective, the above-mentioned limitations will have to be overcome.

Particle bombardment has been very successful for obtaining transgenic plants in cereals such as wheat, maize, and rice and in trees of economic importance such as *Picea glauca* (Table 1). The main advantage of particle bombardment-mediated DNA delivery to obtain transgenic plants is that it is genotype independent as has been shown for rice[49] and cotton.[106] This has been a major limitation of other techniques for DNA delivery into plant cells. The only limitation in producing transgenic plants using this technology is the availability of suitable regeneration protocols that are also amenable to selection of transformed cells using a selective agent.

Particle bombardment has also been very useful in advancing our understanding of the gene regulation and expression as demonstrated by the studies with anthocyanin biosynthesis and regulation in maize. As the technology develops, more plants will be amenable to genetic transformation, allowing gene regulation and expression studies to be performed *in vivo* in tissues of interest. These developments, along with advancements in technologies for

Table 1 *List of transgenic plants produced using biolistics*

Plant Sp.	Gene(s) Introduced	Ref.
Zea mays	*bar, als, gus, lux, hpt*	46, 47, 48
Oryza sativa	*bar, gus, hpt*	49, 50, 51
Triticum aestivum	*bar, gus, nptII*	53, 54
Avena sativa	*bar, gus*	57
Saccharum spp.	*gus, nptII, hpt*	58
Carica papaya	*nptII, gus,* PRV coat protein	61
Gossypium hirsutum	*gus, hpt,* several novel genes	55, 106
Glycine max	*gus, nptII*	27, 28
Nicotiana tabacum	*gus, nptII*	59
Phaseolus vulgaris	*gus, bar*	26
Picea glauca	*gus, nptII, cryIA*	38, 39
Populus sp.	*gus, nptII, bt*	56

producing transgenic plants, will make genetically engineered crops with useful traits a reality in the near future.

Another important use of particle bombardment is its ability to deliver DNA in all three genomes (nucleus, chloroplast, and mitochondria) in the plant cell. The transformation of the mitochondrial genome[107] in yeast and the chloroplast genome[108] in *Chalamydomonas reinharditi* and plants has been demonstrated. Maliga[109] has recently reviewed transformation of the chloroplast genome in plants and has highlighted the role of particle bombardment in chloroplast transformation. The chloroplast genome is present in several copies in a plant cell thus increasing the number of potential targets for particle bombardment-mediated DNA delivery. A homoplastome (cell with all transformed chloroplasts) can be produced by selection of the transformed chloroplast genome. The use of a chimeric *aadA* gene encoding aminoglycoside-3′-adenyltransferase for selection yields a 100-fold increase in transformation frequency as compared to that obtained with spectinomycin resistance encoded by mutant 16S rRNA genes.[110] This makes plastid transformation a practical tool in higher plants. Plastids are the sites of many important biosynthetic pathways for macromolecules such as starch, amino acids, and lipids; however, most of the genes encoding different steps in these pathways are nuclear in origin.[111] Plastid transformation will allow the insertion and expression of gene(s) encoding key step(s) in the plastid and allow the exploration and alteration of these pathways.

In conclusion, particle bombardment has been a successful method of gene delivery, especially in plant species that were not amenable to any other gene delivery method. The utility of particle bombardment in producing transgenic plants is well documented in the literature. As more genes encoding useful traits are identified and characterized, production of genetically transformed plants with desired characteristics will become routine. In important crops such as wheat and corn and other cereals, particle bombardment will be an immensely useful tool.

Acknowledgments

Ms. Karen Caswell is gratefully acknowledged for her help in preparation of this manuscript. Drs. N.S. Nehra, R.S.S. Datla, M.C. Jordan, and V.R. Bommineni are acknowledged for their critical review of this manuscript. Dr. J.C. Sanford of Cornell University is acknowledged for kindly providing the pre-print of Reference 19. This is a National Research Council of Canada publication (NRCC No. 36491).

References

1. **Lindsey, K.,** Genetic manipulation of crop plants, *J. Biotechnol.,* 26, 1, 1992.
2. **Jain, H. K.,** Plant genetic resources and policy, *Trends Biotechnol.,* 6, 73, 1988.
3. **Sanford, J. C., Klein, T. M., Wolf, E. D., and Allen, N.,** Delivery of substances into cells and tissues using a particle bombardment process, *Particulate Sci. Technol.,* 5, 27, 1987.
4. **Sanford, J. C., Wolf, E. D., and Allen, N. K.,** U.S. Patent Appl. 670, 771; U.S. Patent 4,945,050, 1990.
5. **Klein, T. M., Wolf, E. D., Wu, R., and Sanford, J. C.,** High-velocity microprojectiles for delivering nucleic acids into living cells, *Nature,* 327, 70, 1987.
6. **Sanford, J. C.,** The biolistic process, *Trends Biotechnol.,* 6, 299, 1988.
7. **Zumbrunn, G., Schneider, M., and Rochaix, J.-D.,** A simple particle gun for DNA-mediated cell transformation, *Technique,* 1, 204, 1989.
8. **Sanford, J. C., Devit, M. J., Russel, J. A., Smith, F. D., Harpending, P. R., Roy, M. K., and Johnston, S. A.,** An improved, helium-driven biolistic device, *Technique,* 3, 2, 1991.
9. **Christou, P., McCabe, D. E., and Swain, W. F.,** Stable transformation of soybean callus by DNA-coated gold particles, *Plant Physiol.,* 87, 671, 1988.
10. **Morikawa, H., Iida, A., and Yamada, Y.,** Transient expression of foreign genes in plant cells and tissues obtained by a simple biolistic device (particle gun), *Appl. Microbiol. Biotechnol.,* 31, 320, 1989.

11. **Iida, A., Seki, M., Kamada, M., Yamada, Y., and Morikawa, H.,** Gene delivery into cultured plant cells by DNA-coated gold particles accelerated by a pneumatic particle gun, *Theor. Appl. Genet.,* 80, 813, 1990.

12. **Oard, J. H., Paige, D. F., Simmonds, J. A., and Gradziel, T. M.,** Transient gene expression in maize, rice, and wheat cells using an air gun apparatus, *Plant Physiol.,* 92, 334, 1990.

13. **Finer, J. J., Vain, P., Jones, M. W., and McMullen, M. D.,** Development of the particle inflow gun for DNA delivery to plant cells, *Plant Cell Rep.,* 11, 323, 1992.

14. **Sautter, C., Waldner, H., Neuhaus-Url, G., Galli, A., Neuhaus, G., and Potrykus, I.,** Micro-targeting: high efficiency gene transfer using a novel approach for the acceleration of micro-projectiles, *Bio/Technology,* 9, 1080, 1991.

15. **Klein, T. M., Gradziel, T., Fromm, M. E., and Sanford, J. C.,** Factors influencing gene delivery into *Zea mays* cells by high-velocity microprojectiles, *Bio/Technology,* 6, 559, 1988.

16. **Klein, T. M., Arentzen, R., Lewis, P. A., and Fitzpatrick-McElligot, S.,** Transformation of microbes, plants and animals by particle bombardment, *Bio/Technology,* 10, 286, 1992.

17. **Franks, T. M. and Birch, R. G.,** Microprojectile techniques for direct gene transfer into intact plant cells, *Advanced Methods in Plant Breeding and Biotechnology,* Murray, D. R., Ed., C.A.B. International, Wallingford, Oxon, UK, 1992, chap. 5.

18. **Morrish, F. M., Songstad, D. D., Armstrong, C. L., and Fromm, M. E.,** Microprojectile bombardment: a method for the production of transgenic cereal crop plants and the functional analysis of genes, *Transgenic Plants, Fundamentals and Applications,* Hiatt, A., Ed., Marcel Dekker, New York, 1993, chap. 8.

19. **Sanford, J. C., Smith, F. D., and Russel, J. A.,** Optimizing the biolistics process for different biological applications, *Methods Enzymol.,* 217, 483, 1993.

20. **Russel, J. A., Roy, M. K., and Sanford, J. C.,** Physical trauma and tungsten toxicity reduce the efficiency of biolistic transformation, *Plant Physiol.,* 98, 1050, 1992.

21. **Kartha, K. K., Chibbar, R. N., Georges, F., Leung, N., Caswell, K., Kendall, E., and Qureshi, J. A.,** Transient expression of chloramphenicol acetyl transferase (CAT) gene in barley cell cultures and immature zygotic embryos through microprojectile bombardment, *Plant Cell Rep.,* 8, 429, 1989.

22. **Chibbar, R. N., Kartha, K. K., Leung, N., Qureshi, J., and Caswell, K.,** Transient expression of marker genes in immature zygotic embryos of spring wheat (*Triticum aestivum* L.) through microprojectile bombardment, *Genome,* 34, 453, 1991.

23. **Chibbar, R. N., Nehra, N. S., Baga, M., Leung, N., Caswell, K., Steinhauer, L., Mallard, C. S., and Kartha, K. K.,** Stable transformation of wheat callus following microprojectile bombardment (ms submitted).

24. **Klein, T. M., Fromm, M. E., Weissinger, A., Tomes, D., Schaff, S., Sletten, M., and Sanford, J. C.,** Transfer of foreign genes into intact maize cells with high-velocity microprojectiles, *Proc. Natl. Acad. Sci. U.S.A.,* 85, 4305, 1988.

25. **Perl, A., Kless, H., Blumenthal, A., Galili, G., and Galun, E.,** Improvement of plant regeneration and GUS expression in scutellar wheat calli by optimization of culture conditions and DNA-microprojectile delivery procedures, *Mol. Gen. Genet.,* 235, 279, 1992.

26. **Russel, D. R., Wallace, K. M., Bathe, J. M., Martinell, B. J., and McCabe, D. E.,** Stable transformation of *Phaseolous vulgaris* via electric discharge mediated particle acceleration, *Plant Cell Rep.,* 12, 165, 1993.

27. **Christou, P., Swain, W. F., Yang, N.-S., and McCabe, D. E.,** Inheritance and expression of foreign genes in transgenic soybean plants, *Proc. Natl. Acad. Sci. U.S.A.,* 86, 7500, 1989.

28. **McCabe, D. E., Swain, W. F., Martinell, B. J., and Christou, P.,** Stable transformation of soybean *(Glycine max)* by particle acceleration, *Bio/Technology,* 6, 923, 1988.

29. **Franche, C., Bogusz, D., Schopke, C., Fauquet, C., and Beachy, R. G.,** Transient gene expression in cassava using high-velocity microprojectiles, *Plant Mol. Biol.,* 17, 493, 1991.

30. **Prakash, C. S. and Varadarajan, U.,** Genetic transformation of sweet potato by particle bombardment, *Plant Cell Rep.,* 11, 53, 1992.

31. **Warkentin, T. D., Jordan, M. J., and Hobbs, S. L. A.,** Effect of promoter sequences on transient reporter gene expression in particle bombarded pea (*Pisum sativum* L.) tissues, *Plant Sci.,* 87, 171, 1992.

32. **Fitch, M. M. M.., Manshardt, R. M., Gonsalves, D., Slightom, J. L., and Sanford, J. C.,** Stable transformation of papaya via microprojectile bombardment, *Plant Cell Rep.,* 9, 189, 1990.

33. **Genga, A., Ceriotti, A., Bollini, R., Bernacchia, G., and Allavena, A.,** Transient gene expression in bean tissues by high-velocity microprojectile bombardment, *J. Genet. Breed.,* 45, 129, 1991.

34. **Lonsdale, D., Onde, S., and Cuming, A.,** Transient expression of exogenous DNA in intact viable wheat embryos following particle bombardment, *J. Exp. Bot.,* 41, 1161, 1990.

35. **Franks, T. and Birch, R. G.,** Gene transfer into intact sugarcane cells using microprojectile bombardment, *Aust. J. Plant Physiol.*, 18, 471, 1991.

36. **Wang, Y.-C., Klein, T. M., Fromm, M. E., Cao, J., Sanford, J. C., and Wu, R.,** Transient expression of foreign genes in rice, wheat and soybean cells following particle bombardment, *Plant Mol. Biol.*, 11, 433, 1988.

37. **Mendel, R. R., Muller, B., Schulze, J., Kolensikov, V., and Zelenin, A.,** Delivery of foreign genes to intact barley cells by high-velocity microprojectiles, *Theor. Appl. Genet.*, 78, 31, 1989.

38. **Ellis, D. D., McCabe, D. E., Mcinnis, S., Ramachandran, R., Russel, D. R., Wallace, K. M., Martinell, B. J., Roberts, D. R., Raffa, K. F., and McCown, B. H.,** Stable transformation of *Picea glauca* by particle acceleration, *Bio/ Technology*, 11, 84, 1993.

39. **Bommineni, V. R., Chibbar, R. N., Datla, R. S. S., and Tsang, E. W. T.,** Transformation of white spruce *(Picea glauca)* somatic embryos by microprojectile bombardment, *Plant Cell Rep.*, 13, 17, 1993.

40. **Yamashita, T., Iida, A., and Morikawa, H.,** Evidence that more than 90% of β-glucuronidase-expressing cells after particle bombardment directly receive the foreign gene in their nucleus, *Plant Physiol.*, 97, 829, 1991.

41. **Taylor, M. K. and Vasil, I. K.,** Histology of, and physical factors affecting, transient GUS expression in pearl millet (*Pennisetum glaucum* (L.) R.Br.) embryos following microprojectile bombardment, *Plant Cell Rep.*, 10, 120, 1991.

42. **Klein, T. M., Harper, E. C., Svab, Z., Sanford, J. C., Fromm, M. E., and Maliga, P.,** Stable genetic transformation of intact *Nicotiana* cells by the particle bombardment process, *Proc. Natl. Acad. Sci. U.S.A.*, 85, 8502, 1988.

43. **Klein, T. M., Kornstein, L., Sanford, J. C., and Fromm, M. E.,** Genetic transformation of maize cells by particle bombardment, *Plant Physiol.*, 91, 440, 1989.

44. **Vasil, V., Brown, S. M., Re, D., Fromm, M. E., and Vasil, I. K.,** Stably transformed callus lines from microprojectile bombardment of cell suspension cultures of wheat, *Bio/Technology*, 9, 743, 1991.

45. **Hagio, T., Blowers, A. D., and Earle, E. D.,** Stable transformation of sorghum cell cultures after bombardment with DNA-coated microprojectiles, *Plant Cell Rep.*, 10, 260, 1991.

46. **Fromm, M. E., Morrish, F., Armstrong, C., Williams, R., Thomas, J., and Klein, T. M.,** Inheritance and expression of chimeric genes in the progeny of transgenic maize plants, *Bio/Technology*, 8, 833, 1990.

47. **Gordon-Kamm, W. J., Spencer, T. M., Mangano, M. L., Adams, T. R., Daines, R. J., Start, W. G., O'Brien, J. V., Chambers, S. A., Adams, W. R., Jr., Willets, N. G., Rice, T. B., Mackey, C. J., Kruger, R. W., Kausch, A. P., and Lemaux, P. G.,** Transformation of maize cells and regeneration of fertile transgenic plants, *Plant Cell,* 2, 603, 1990.

48. **Walters, D. A., Vetsch, C. S., Potts, D. E., and Lundquist, R. C.,** Transformation and inheritance of a hygromycin phosphotransferase gene in maize plants, *Plant Mol. Biol.,* 18, 189, 1992.

49. **Christou, P. and Ford, T. L.,** Production of transgenic rice (*Oryza sativa* L.) plants from agronomically important indica and japonica varieties via electric discharge particle acceleration of exogenous DNA into immature zygotic embryos, *Bio/Technology,* 9, 957, 1991.

50. **Cao, J., Duan, X., McElroy, D., and Wu, R.,** Regeneration of herbicide resistant transgenic rice plants following microprojectile-mediated transformation of suspension culture cells, *Plant Cell Rep.,* 11, 586, 1992.

51. **Li, L., Qu, R., de Kochko, A., Fauquet, C., and Beachy, R. N.,** An improved rice transformation system using the biolistic method, *Plant Cell Rep.,* 12, 250, 1993.

52. **Kartha, K. K., Chibbar, R. N., Nehra, N. S., Leung, N., Caswell, K., Baga, M., Mallard, C. S., and Steinhauer, L.,** Genetic engineering of wheat through microprojectile bombardment using immature zygotic embryos, *J. Cell. Biochem.,* 16F, 198, 1992

53. **Vasil, V., Castillo, A. M., Fromm, M. E., and Vasil, I. K.,** Herbicide resistant fertile transgenic wheat plants obtained by microprojectile bombardment of regenerable embryogenic callus, *Bio/Technology,* 10, 667, 1992.

54. **Nehra, N. S., Chibbar, R. N., Leung, N., Caswell, K., Mallard, C., Steinhauer, L., Baga, M., and Kartha, K. K.,** Self-fertile transgenic wheat plants regenerated from isolated scutellar tissues following microprojectile bombardment with two distinct gene constructs, *Plant J.,* 5, 285, 1994.

55. **Finer, J. J. and McMullen, M. D.,** Transformation of cotton (*Gossypium hirsutum* L.) via particle bombardment, *Plant Cell Rep.,* 8, 586, 1990.

56. **McCown, B. H., McCabe, D. E., Russel, D. R., Robison, D. J., Barton, K. A., and Raffa, K. F.,** Stable transformation of *Populus* and incorporation of pest resistance by electric discharge particle acceleration, *Plant Cell Rep.,* 9, 590, 1991.

57. **Somers, D. A., Rines, H. W., Gu, W., Kaeppler, H. F., and Bushnell, W. R.,** Fertile, transgenic oat plants, *Bio/Technology,* 10, 1589, 1992.

58. **Bower, R. and Birch, R. G.,** Transgenic sugarcane plants via microprojectile bombardment, *Plant J.*, 2, 409, 1992.

59. **Tomes, D. T., Weissinger, A. K., Ross, M., Higgins, R., Drummond, B. J., Schaff, S., Malone-Schoneberg, J., Stabell, M., Flynn, P., Anderson, J., and Howard, J.,** Transgenic tobacco plants and their progeny derived by microprojectile bombardment of tobacco leaves, *Plant Mol. Biol.*, 14, 261, 1990.

60. **Seiki, M., Shigemoto, N., Komeda, Y., Imamura, J., Yamada, Y., and Morikawa, H.,** Transgenic *Arabidopsis thaliana* plants obtained by particle-bombardment-mediated transformation, *Appl. Microbiol. Biotechnol.*, 36, 228, 1991.

61. **Fitch, M. M. M., Manshard, R. M., Gonsalves, D., Slightom, J. L., and Sanford, J. C.,** Virus resistant papaya plants derived from tissues bombarded with the coat protein gene of papaya ring spot virus, *Bio/Technology*, 10, 1466, 1992.

62. **Kuenhle, A. R. and Sugii, N.,** Transformation of *Dendrobium* orchid using particle bombardment of protocorms, *Plant Cell Rep.*, 11, 484, 1992.

63. **Hauptmann, R. M., Vasil, V., Ozias-Akins, P., Tabaeizadeh, Z., Rogers, S. G., Fraley, R. T., Horsch, R. B., and Vasil, I. K.,** Evaluation of selectable markers for obtaining stable transformants in the gramineae, *Plant Physiol.*, 86, 602, 1988.

65. **Fraley, R. T., Rogers, S. G., Horsch, R. B., Eicholtz, D. A., Flick, J. S., Adams, S. P., Bittner, M. L., Brans, L. A., Fink, C. L., Fry, J. S., Galluppi, G. R., Goldberg, S. B., Hoffman, N. L., and Woo, S. C.,** Expression of bacterial genes in plant cells, *Proc. Natl. Acad. Sci. U.S.A.*, 80, 4803, 1983.

66. **Depicker, A., Stachel, S., Shase, P., Zambryski, P., and Goodman, H. M.,** Nopaline synthase: transcript mapping and DNA sequence, *J. Mol. Appl. Genet.*, 1, 561, 1982.

67. **Velten, J., Velten, L., Hain, R., and Schell, J.,** Isolation of a dual promoter fragment from the Ti plasmid of *Agrobacterium tumefaciens, EMBO J.*, 3, 2723, 1984.

68. **Guilley, H., Dudley, R. K., Jonard, G., Balazas, E., and Richards, K. E.,** Transcription of cauliflower mosaic virus DNA: detection of promoter sequences and characterization of transcripts, *Cell,* 30, 763, 1982.

69. **Odell, J. T., Nagy, F., and Chua, N.-H.,** Identification of DNA sequences required for the activity of the cauliflower mosaic virus 35S promoter, *Nature,* 313, 810, 1985.

70. **Fluhr, R., Kuhlemeyer, C., Nagy, T. F., and Chua, N.-H.,** Organ-specific and light induced expression of plant genes, *Science,* 232, 1106, 1986.

71. **Fang, R.-X., Nagy, F., Sivasubramaniam, S., and Chua, N.-H.,** Multiple *cis* regulatory elements for maximal expression of the cauliflower mosaic virus 35S promoter in transgenic plants, *Plant Cell,* 1, 141, 1989.

72. **Kay, R., Chan, A., Daly, M., and McPherson, J.,** Duplication of CaMV35S promoter sequences creates a strong enhancer for plant genes, *Science,* 236, 1299, 1987.

73. **Benfey, P. N. and Chua, N.-H.,** Regulated genes in transgenic plants, *Science,* 244, 174, 1989.

74. **Chibbar, R. N., Kartha, K. K., Datla, R. S. S., Leung, N., Caswell, K., Mallard, C. S., and Steinhauer, L.,** The effect of different promoter-sequences on transient expression of *gus* reporter gene in cultured barley (*Hordeum vulgare* L.) cells, *Plant Cell Rep.,* 12, 506, 1993.

75. **Dennis, E. S., Gerlach, W. L., Pryor, A. J., Bennetzen, J. L., Inglis, A., Ilewellyn, D., Sachs, M. M., Ferl, R. J., and Peacock, W. J.,** Molecular analysis of the alcohol dehydrogenase *(Adh1)* gene of maize, *Nucleic Acids Res.,* 12, 3983, 1984.

76. **Walker, J. C., Howard, E. A., Dennis, E. S., and Peacock, W. J.,** DNA sequences required for the anaerobic expression of the maize alcohol dehydrogenase 1 gene, *Proc. Natl. Acad. Sci.*'*U.S.A.,* 84, 6624, 1987.

77. **Olive, M. R., Walker, J. C., Singh, K., Dennis, E. S., and Peacock, W. J.,** Functional properties of the anaerobic responsive element of the maize *Adh1* gene, *Plant Mol. Biol.,* 15, 593, 1990.

78. **Last, D. I., Brettel, R. I. S., Chamberlain, D. A., Chaudhary, A. M., Larkin, P. J., Marsh, E. L., Peacock, W. J., and Dennis, E. S.,** pEMU: an improved promoter for gene expression in cereal cells, *Theor. Appl. Genet.,* 81, 581, 1991.

79. **Ellis, J. G., Llewellyn, D. J., Walker, J. C., Dennis, E. S., and Peacock, W. J.,** The *ocs* element: a 16 base pair palindrome essential for activity of the octopine synthase enhancer, *EMBO J.,* 6, 3203, 1987.

80. **McElroy, D., Zhang, W., Cao, J., and Wu, R.,** Isolation of an efficient actin promoter for use in rice transformation, *Plant Cell,* 2, 163, 1990.

81. **McElroy, D., Blowers, A. D., Jenes, B., and Wu, R.,** Construction of expression vectors based on the rice actin 1 *(Act1)* 5′ region for use in monocot transformation, *Mol. Gen. Genet.,* 231, 150, 1991.

82. **Christensen, A. H., Sharrock, R. A., and Quail, P. H.,** Maize polyubiquitin genes: structure, thermal perturbation of expression and transcript splicing, and promoter activity following transfer to protoplasts by electroporation, *Plant Mol. Biol.,* 18, 675, 1992.

83. **de Pater, B. S., van der Mark, F., Rueb, S., Katagiri, F., Chua, N.-H., Schilerpoort, R. A., and Hensgens, A. M.,** The promoter of the rice gene *GOS2* is active in various monocot tissues and binds rice nuclear factor ASF-1, m, *Plant J.,* 2, 837, 1992.

84. **Gallie, D. R., Sleat, D. E., Watts, J. W., Turner, P. C., and Wilson, T. M. A.,** The 5′-leader sequence of tobacco mosaic virus RNA enhances the expression of foreign gene transcripts *in vitro* and *in vivo, Nucleic Acids Res.,* 15, 3257, 1987.

86. **Nicolaisen, N., Johansen, E., Poulsen, G. B., and Borkhardt, B.,** The 5′ untranslated region from pea seedborne mosaic potyvirus RNA as a translational enhancer in pea and tobacco protoplasts, *FEBS,* 303, 169, 1992.

87. **Leon, P., Planckaert, F., and Walbot, V.,** Transient gene expression in protoplasts of *Phaseolus vulgaris* isolated from a cell suspension culture, *Plant Physiol.,* 95, 968, 1991.

88. **Callis, J., Fromm, M., and Walbot, V.,** Introns increase gene expression in cultured maize cells, *Genes Dev.,* 1, 1183, 1987.

89. **Mascarenhas, D., Mettler, I. J., Pierce, D. A., and Lowe, H. W.,** Intron-mediated enhancement of heterologous gene expression in maize, *Plant Mol. Biol.,* 15, 913, 1990.

90. **Vasil, V., Clancy, M., Ferl, R. J., Vasil, I. K., and Hannah, L. C.,** Increased gene expression by the first intron of maize *shrunken-1* locus in grass species, *Plant Physiol.,* 91, 1575, 1989.

91. **Maas, C., Laufs, J., Grant, S., Korfhage, C., and Werr, W.,** The combination of a novel stimulatory element in the first exon of maize *shrunken-1* gene with the following intron 1 enhances reporter gene expression up to 1000-fold, *Plant Mol. Biol.,* 16, 199, 1991.

92. **Luehrsen, K. R. and Walbot, V.,** Intron enhancement of gene expression and the splicing efficiency of introns in maize cells, *Mol. Gen. Genet.,* 225, 81, 1991.

93. **Siva, E. M., Mettler, I. J., Dietrich, P. S., and Sinbaldi, R. M.,** Enhanced transient expression in maize protoplasts, *Genome,* 30, A72, 1988.

94. **Tanaka, A., Mita, S., Ohta, S., Kyozuka, J., Shimamoto, K., and Nakamura, K.,** Enhancement of foreign gene expression by a dicot intron in rice but not in tobacco is correlated with an increased level of mRNA and an efficient splicing of the intron, *Nucleic Acids Res.,* 18, 6767, 1991.

95. **Vernet, T., Fleck, J., Durr, A., Fritsch, C., Pinck, M., and Hirth, L.,** Expression of the gene coding for the small subunit of ribulosebiphosphate carboxylase during differentiation of tobacco plant protoplasts, *Eur. J. Biochem.,* 126, 489, 1982.

96. **Spencer, T. M., O'Brien, J. V., Start, W. G., Adams, T. R., Gordon-Kamm, W. J., and Lemaux, P. G.,** Segregation of transgenes in maize, *Plant Mol. Biol.,* 18, 201, 1992.

97. **Ingelbrecht, I. L. W., Herman, L. M. F., Dekeyser, R. A., Van Montagu, M., and Depicker, A. G.,** Different 3′ end regions strongly influence the level of gene expression in plant cells, *Plant Cell,* 1, 671, 1989.

98. **Dekeyser, R. A., Claes, B., De Rycke, R. M. U., Habets, M. E., Van Montagu, M., and Caplan, A. B.,** Transient gene expression in intact and organized rice tissue, *Plant Cell,* 2, 591, 1990.

99. **Klein, T. M., Roth, B. A., and Fromm, M. E.,** Regulation of anthocyanin biosynthetic genes introduced into intact maize tissues by microprojectiles, *Proc. Natl. Acad. Sci. U.S.A.,* 86, 6681, 1989.

100. **Goff, S. A., Klein, T. M., Roth, B. A., Fromm, M. E., Cone, K. C., Radicella, J. P., and Chandler, V. L.,** Transactivation of anthocyanin biosynthetic genes following transfer of *B* regulatory genes into maize tissues, *EMBO J.,* 9, 2517, 1990.

101. **Ludwig, S. R., Bowen, B., Beach, L., and Wessler, S. R.,** A regulatory gene as a novel visible marker for maize transformation, *Science,* 247, 449, 1990.

102. **Coe, E. H., Neuffer, M. G., and Hoisington, D. A.,** The genetics of corn, *Corn and Corn Improvement,* Spague, G. F. and Dudley, J., Eds., American Agronomy Society, Madison, WI, 1988, 81.

103. **Bruce, W. B., Christensen, A. H., Klein, T. M., Fromm, M., and Quail, P. H.,** Photoregulation of a phytochrome gene promoter from oat transferred into rice by particle bombardment, *Proc. Natl. Acad. Sci. U.S.A.,* 86, 9692, 1989.

104. **Knudsen, S. and Muller, M.,** Transformation of the developing barley endosperm by particle bombardment, *Planta,* 185, 330, 1991.

105. **Kim, J.-K., Cao, J., and Wu, R.,** Regulation and interaction of multiple protein factors with the proximal promoter regions of a rice high pI α-amylase gene, *Mol. Gen. Genet.,* 232, 383, 1992.

106. **McCabe, D. E. and Martinell, B. J.,** Transformation of elite cotton cultivars via particle bombardment of meristems, *Bio/Technology,* 11, 596, 1993.

107. **Johnston, S. A., Anziano, P. Q., Shark, K., Sanford, J. C., and Butow, R. A.,** Mitochondrial transformation in yeast by bombardment with microprojectiles, *Science,* 240, 1538, 1988.

108. **Boynton, J. E., Gillham, N. W., Harris, E. H., Hosler, J. P., Johnson, A. M., Jones, A. R., Randolph-Anderson, B. L., Robertson, D., Klein, T. M., Shark, K. B., and Sanford, J. C.,** Chloroplast transformation in *Chlamydomonas* with high velocity microprojectiles, *Science,* 240, 1538, 1988.

109. **Maliga, P.,** Towards plastid transformation in flowering plants, *Trends Biotechnol.,* 11, 101, 1993.

110. **Svab, Z. and Maliga, P.,** High-frequency plastid transformation in tobacco by selection for a chimeric *aadA* gene, *Proc. Natl. Acad. Sci. U.S.A.,* 90, 913, 1993.

111. **Mullet, J. E.,** Chloroplast development and gene expression, *Annu. Rev. Plant Physiol. Mol. Biol.,* 39, 475, 1988.

The Transformation of Legumes Using *Agrobacterium tumefaciens*

Mark C. Jordan[1] and Shaun L. A. Hobbs[2]

[1] *National Research Council of Canada, Plant Biotechnology Institute, Saskatoon, Saskatchewan, Canada*

[2] *Department of Plant Breeding and Genetics, CAB International, Wallingford, Oxfordshire, United Kingdom*

Introduction

Agrobacterium tumefaciens is a Gram-negative soil bacterium that causes crown gall disease in many plants. Wild-type strains have the ability to introduce a segment of their own DNA, the transfer- (T-) DNA, into the plant cell. There the genes on the T-DNA are integrated into the plant genome and expressed, producing opines (carbon and nitrogen sources for bacterial growth) and hormones that cause an unregulated proliferation of callus tissue.[1,2] The T-DNA is a specific segment on the tumor-inducing plasmid of *A. tumefaciens* and is delimited by borders of repeated sequences. The naturally occurring T-DNA can be replaced, through molecular techniques, with genes from any organism, and these genes can then be introduced into the plant genome using *A. tumefaciens* as a delivery system.[3] Once integrated into the plant's DNA, the alien genes can be passed on to future generations in a stable Mendelian fashion, and if these genes are fused to promoters which are recognized by the plant, then they will be predictably expressed in their new host. Using this

0-8493-8262-9/94/$0.00+$.50

method, genes for economically valuable traits can now be introduced into many crop plants.[4–6]

The Leguminosae family provides a variety of important crops that are grown in temperate, subtropical, and tropical parts of the world.[7,8] Grain or pulse legumes used as cash crops, animal feed, or staple human foods include soybean (*Glycine max* Merr.), groundnut (*Arachis hypogaea* L.), dry bean (*Phaseolus vulgaris* L.), pea (*Pisum sativum* L.), broad bean (*Vicia faba* L.), cowpea (*Vigna unguiculata* L.), and pigeonpea (*Cajanus cajun* L.). Important forage legumes, which provide fodder for many types of domestic animals, include alfalfa (*Medicago sativa* L.), clovers (*Trifolium* spp.), birdsfoot trefoil (*Lotus corniculatus* L.), and *Stylosanthes* spp. Legumes have high energy and protein contents which make them particularly nutritionally valuable. In addition, they have an ability to enter into a symbiotic relationship with nitrogen fixing bacteria, allowing them to grow in, and enrich, soil with poor fertility. Therefore, the use of transformation methodology for improvement of this important plant group has great economic potential.

The basic stages in the development of an *Agrobacterium*-mediated transformation system for any crop are: (1) the development of an *in vitro* regeneration system whereby plants can be regenerated from a cell or small group of cells; (2) the determination that these cells are susceptible to transformation by *Agrobacterium;* (3) the isolation and introduction of selectable markers and reporter genes that allow the transformed plant material to be distinguished; and (4) the successful regeneration of transformed material, producing plants that express the introduced genes and pass them on to their offspring. These stages will be discussed here in relation to the success and problems met in transforming legumes.

Regeneration

The production of transformed plants involves the stable insertion of T-DNA into a cell, or small number of cells, that can then regenerate into a whole plant. A regeneration system must generally meet several criteria if it is to be useful for transformation. The regenerating tissue should ideally be at, or close to, the surface of an explant so that it is easily accessible to the A*grobacterium* during co-cultivation. In addition, as successful transformation is a comparatively rare event, the regenerating system should be rapid and have the potential to produce many regenerants per explant. The fewer the number of cells producing the regenerated plant the better. If regeneration occurs from a single transformed cell, then all the cells in the resulting plant will include the alien gene, ensuring that the gene will be passed on to any offspring. However, if many cells produce the regenerant and only a few are transformed, then a chimera is produced and the transgene will only be stably inherited if the germ cells are produced from the transformed part of the plant. It is also desirable that any regeneration system work on as many of the economically valuable genotypes as possible, and it is imperative that the cell culture process does not induce infertility or unintentional variability, through somaclonal variation, in the regenerants.

It has been difficult in the past to find regeneration methods for legumes, particularly grain legumes, that fit these criteria. However, methods of regeneration through organogenesis, somatic embryogenesis, and protoplast manipulation have recently been developed that should be generally amenable to use in transforming important legume species (as reviewed by Bajaj[9] and Mroginski and Kartha[10]).

Susceptibility of Legumes to *Agrobacterium*

Wild-type *A. tumefaciens* does not attack all plants, and it is important to establish that the *Agrobacterium* strain to be used and the species and genotype to be transformed are compatible. Wild-type strains have been found to be virulent on soybean,[11–15] pea,[16–18] dry bean,[19,20] lentil (*Lens culinaris* Medik.),[21] moth bean [*Vigna aconitifolia* (Jacq.) Marechal],[22] and various forage legumes.[11,13,23–25] However, there is considerable variation among the legume species, as well as in the ability of the different strains to infect legumes[15,16,18,20,21,23,25] and in the response of the different genotypes within a species to that infection.[12,15–18,20,24,25] Such interactions can be very important as, depending on the strain-genotype combinations used, soybean[12] and alfalfa[25] ranged in response from nonsusceptible to very highly susceptible. Southern hybridization has been used to show that the T-DNA can be transferred to and integrated into legume genomic DNA,[12–14,18,21,25] and confirmation of the functioning of the introduced genes has been produced by the detection of opines in callusing tissue.[12,14,18,21,23,25]

Transfer of Selectable Marker and Reporter Genes into Legume Tissues

Disarmed strains of *Agrobacterium,* i.e., those where native genes on the T-DNA have been replaced by other genes, can be co-cultivated with explants to introduce many different scorable and selectable marker genes into plant tissues. Genes that code for antibiotic resistance (e.g., neomycin phosphotransferase for kanamycin,[26] hygromycin phosphotransferase for hygromycin[27]) or herbicide resistance (e.g., 5-enolpyruvylshikimate-3-phosphate synthase for glyphosate,[28] phosphinothricin acetyltransferase for phosphinothricin and bialaphos[29]) provide mechanisms by which transformed tissue can be selected. When expressed in the plant tissue they confer the ability for transformed material to survive on media containing levels of the toxic substance that would kill untransformed material. In addition, reporter genes, such as those coding for nopaline synthase[30] or β-glucuronidase (GUS),[31] can be introduced into the plant, where their expression can quickly and easily identify transformed tissue. In addition, reporter genes can identify the cells or cell types within the tissue that are expressing the gene.

Differences in co-cultivation conditions of *Agrobacterium* and explant and subsequent methods of selection have been found to have a marked effect on the recovery of transgenic plants in many species (e.g., tomato[32] and alfalfa[33]). Using selectable and scorable marker genes, the best methods of co-cultivating *Agrobacterium* and legume tissue explants can be devised. These can be determined on regenerating tissue but can be monitored more rapidly on callusing systems. Hence, the best conditions for producing transgenic callus were determined for soybean, and these conditions were then used to produce transgenic plants.[34] Many different factors have now been reported to affect the efficiency of legume transformation during co-cultivation and selection. These include co-cultivation time,[33,35–37] presence of nurse cells[35] or inducing compounds such as acetosyringone,[38] type of strains,[33,35,36,39] and type of selection medium.[35,37] By optimizing these factors, the potential effectiveness of *Agrobacterium*-mediated transformation of legumes has been clearly demonstrated through the production of transformed callus in many forage and grain legume species.[15,16,20,33,35–37,39–42]

Production of Transgenic Legumes

Problems

To be useful for crop improvement, a transformation system must rapidly produce large numbers of transgenic plants, have little or no callus stage to reduce somaclonal variation, and be genotype independent so that elite genotypes can be used. Although a number of *A. tumefaciens*-mediated transformation systems have resulted in the production of transgenic shoots (Table 1), none approaches the ideal system. Only in alfalfa,[45,46] soybean,[34,53] and *Stylosanthes humilis*[50] can fertile transgenic plants be repeatedly produced. Even in these species, the regeneration procedure is strongly genotype-dependent and requires a callus phase of at least 1 month in soybean and 3 to 4 months for alfalfa and *Stylosanthes*. An *Agrobacterium* strain x genotype interaction also exists for soybean, and some cultivars are resistant to *Agrobacterium*-mediated transformation.[34] Chee et al.[53] inoculated the plumule and cotyledonary node regions of germinating soybean seeds in order to avoid a tissue culture phase. Although chimeric plants were obtained and some of these produced transgenic progeny, the frequency was low with only 0.07% of the inoculated seeds giving rise to plants which transmitted the foreign genes to their progeny.

The major problem in the development of commercially viable *Agrobacterium*-mediated transformation systems is a lack of efficient regeneration systems in which a large proportion of cells in the explant are capable of regeneration. Presently, rapid regeneration systems involve plant regeneration

Table 1 *Summary of reports describing the production of transformed shoots in legume species*

Species	Explant	Promoter gene	Expression[a]	Southern[b]	Progeny[c]	Ref.
Medicago sativa	Stem	*nos*-nptII	Yes	Yes	NR	39
	Stem, petiole	35S-*bar*	Yes	Yes	NR	43
		nos-nptII	Yes			
	Shoot, leaf	*nos-nptII*	Yes	Yes	NR	44
	Leaf	35S-AMV coat protein	Yes	Yes	Yes	45
		19S-*nptII*	Yes			
	Stem	*mas-nptII*	Yes	Yes	Yes	46
Medicago varia	Stem	*nos-nptII*	Yes	Yes	NR	47
	Stem, petiole	*nos-nptII*	Yes	Yes	NR	33
Vigna unguiculata	Embryos	*nos-nptII*	NR	No	No	48
		35S-*uidA*-INT	Yes			
Vigna aconitifolia	Protoplasts	*nos-nptII*	Yes	Yes	NR	49
		nos-nos	Yes			
Stylosanthes humilis	Leaf	*nos-nptII*	Yes	Yes	Yes	50
		nos-nos	Yes			
Vicia narbonensis	Epicotyl	35S-*hpt*	NR	Yes	No	51
	Shoot tip	*nos-nos*	Yes			
Trifolium repens	Stolon	*nos-nptII*	Yes	No	NR	36
	Internodes	*nos-nos*	Yes			
Phaseolus vulgaris coccineus	Stem	*nos-nptII*	NR	No	No	52
		35S-*uidA*	Yes			

Table 1 (continued) *Summary of reports describing the production of transformed shoots in legume species*

Species	Explant	Promoter gene	Expression[a]	Southern[b]	Progeny[c]	Ref.
Glycine max	Cotyledon	*nos-nptII*	Yes	Yes	Yes	34
		35S-*uidA*	Yes			
		35S35S-EPSP synthase	Yes			
	Embryos	*nos-nptII*	Yes	Yes	Yes	53
	Cotyledon	*nos-nptII*	NR	No	No	55
		35S-Ac *uidA*	Yes			
Pisum sativum	Epicotyl	*nos-nptII*	Yes	No	No	37
	Node	35S-*uidA*-INT	Yes			
	Epicotyl	35S-*hpt*	NR	Yes	NR	56
Lotononis bainesii	Leaf	*nos-nptII* 19S-synthetic gene	Yes	No	Yes	57

Note: NR = not reported.

Abbreviations: nos = nopaline synthase; *nptII* = neomycin phosphotransferase (kanamycin resistance); 35S = cauliflower mosaic virus 35S promoter; *bar* = bialaphos resistance (also provides resistance to phosphinothricin and glufosinate ammonium); AMV = alfalfa mosaic virus; 19S = cauliflower mosaic virus 19S promoter; *mas* = mannopine synthase promoter; *uidA* = gene encoding β-glucuronidase; *uidA*-INT = *uidA* gene containing a plant intron to prevent bacterial expression; 35S35S = duplicated 35S promoter; EPSP = 5-enolpyruvylshikimate-3-phosphate (target protein for the herbicide glyphosate); Ac *uidA* = maize controlling element Ac containing the *uidA* gene.

[a] By enzyme assay.

[b] Southern analysis performed on DNA from shoot material.

[c] Foreign gene(s) shown to be present in progeny populations by expression assay or Southern.

from complex explants such as pea cotyledonary nodes[37,58] or soybean cotyledon explants.[34] Only a few cells in these explants are capable of plant regeneration, and these must be competent for transformation as well as accessible to *Agrobacterium* for the production of transgenic plants. A transgenic pea plant has been recovered from a cotyledonary node explant,[77] and transgenic soybean plants can be produced from cotyledon explants;[34] therefore, it is possible to transform regeneration competent cells. The low frequencies observed are perhaps due to the limited number of these cells in the explants, coupled with a low frequency of cellular transformation events. In pea, cotyledonary node explants were examined by the histochemical GUS assay for cellular transformation one week after co-cultivation with *Agrobacterium* which carried an intron containing *uidA* gene encoding GUS. There was an average of only 10 blue staining loci per explant and most of these were not in the area of regeneration.[78]

Further research on the development of improved regeneration systems having a large number of regeneration competent cells in the culture is needed. Reliable embryogenic callus or suspension culture systems would prove to be a major step towards the commercial application of transgenic technology to legume species.[59]

Current Applications

Transgenic plants carrying genes for resistance to broad spectrum herbicides will be among the first applications of genetic engineering to reach the field; in this respect, legumes are no exception. Hinchee et al.[34] produced a soybean plant carrying a gene coding for a resistant form of the target enzyme of the herbicide glyphosate. The gene was inherited by the progeny and expressed. The ability of excised leaves to callus on glyphosate levels which prevented callusing of control leaves was demonstrated; however, no whole plant testing was reported.

Alfalfa plants were transformed with constructs carrying the *bar* gene, which detoxifies the herbicide glufosinate-ammonium.[43] There were 59 transgenic lines produced, and complete herbicide resistance was demonstrated both in the greenhouse and in the field for those lines exhibiting the highest expression of the *bar* gene.

Resistance to diseases caused by fungal, bacterial, or viral pathogens has also been an area of much recent study. In the legumes, resistance to infection by the alfalfa mosaic virus (AMV) has been achieved by transforming alfalfa plants with the gene for the viral coat protein under control of the cauliflower mosaic virus 35S promoter.[45] Transgenic plants containing large amounts of coat protein were resistant to infection with up to 50 μg/ml AMV, while 10 μg/ml was enough to systemically infect control plants.

Future Prospects

A transformation system can also provide new opportunities for the isolation of economically important or scientifically interesting genes. Techniques such as T-DNA insertion mutagenesis and transposon tagging can be used to clone genes into which these foreign DNAs have been inserted.[60] Such techniques have the potential to be useful in the diploid grain legumes such as soybean and pea. Zhou and Atherly[55] examined the possibility of using the transposable element *Ac* from maize for gene tagging in soybean. They found that, although no fertile transformants were obtained, the element was capable of transposition in the leaves, stems, and roots of primary transformants.

A transformation system would also help to increase our knowledge of the function and regulation of isolated legume genes. For example, cDNA sequences for pea genes specifically induced by infection with pea fungal and bacterial pathogens have been isolated.[61] A pea transformation system would allow the manipulation of these genes and their reinsertion in the pea plant to determine their role in disease resistance and to study their regulation.

To date, heterologous systems have been used to study the regulation of legume genes where no transformation system exists for that species. Pea seed storage proteins have been isolated and expressed in tobacco where they are correctly regulated;[62,63] however, these genes would be better studied in a homologous system. This would allow observation of the effect of altering the expression of one or more genes on the expression of other seed storage protein genes, as their regulation is complex and interconnected. Altering the regulation of specific seed storage protein genes may allow a change in the seed protein profile for increased nutritive value. For example, the seeds of grain legumes are low in the essential amino acid methionine.[8] Increasing the proportion of legumin to vicilin while maintaining the same level of total protein would increase the amount of methionine in the seed protein.

Another method for increasing the methionine levels in seed protein has been successful in *Brassica napus* L. A gene encoding a high methionine albumin was isolated from Brazil nut and transferred to *Brassica,* resulting in enhanced levels of methionine in the seed protein.[64] This gene would be a good candidate for insertion into legumes with the aim of improving their nutritional value. Synthetic genes encoding essential amino acids could also help to improve the amino acid profile of legume protein. Wier et al.[57] transformed *Lotononis bainesii* Baker with a gene for a synthetic protein which was high in isoleucine, lysine, methionine, threonine, and tryptophan content, and some of the transgenic plants exhibited higher levels of these amino acids.

Forage legumes could also be open to other nutritional improvement. For example, ruminant animals grazing on clover or alfalfa are susceptible to bloat, and it has been suggested that by cloning the genes for tannin synthesis from *L. corniculatus* and inserting them into clover this problem could be reduced or eliminated.[65]

Legumes are rich in chemical constituents which are considered to be antinutritional factors such as tannins, protease inhibitors, lectins, polyphenols, goitrogens, and cyanogenic glycosides, some of which are destroyed by cooking. By the use of recombinant DNA techniques the genes responsible for these factors could be isolated, modified, and reinserted into the plant in order to improve the nutritional quality and digestibility of the protein. For example, reinsertion of the genes in an antisense orientation may eliminate the production of the antinutritional factor.[66]

However, some of the antinutritional factors, such as lectins and trypsin inhibitors, may be useful as they may play a role in resistance to insect pests whose larvae can do great damage to stored legume seeds. A lectin from a wild species of *P. vulgaris* was found to have toxic effects on an insect pest of cultivated bean, and transfer of the gene to bean cultivars by traditional breeding resulted in insect resistance.[67] The gene encoding this lectin could be isolated and transferred to other legumes via *Agrobacterium*-mediated transformation. A candidate for such a transfer would be cowpea, as lectins from several plant species have been found to be effective against the cowpea weevil.[68] Proteinase inhibitors also can play a role in insect resistance, since transformation of tobacco with genes encoding bean α-amylase inhibitor or cowpea trypsin inhibitor provided increased levels of insect resistance.[69,70] These genes would also be candidates for improving the resistance of legume seeds to insect pests.

Legumes form a symbiotic arrangement with soil bacteria of the Rhizobiaceae family which have the important ability to fix atmospheric nitrogen. The complex processes involved in infection and nodule formation are as yet poorly understood. Transgenic legume plants may provide a model system in which to study the role of the genes involved. *A. rhizogenes* strains carrying root and/or nodule specific genes have been used to incite hairy roots and investigate nodulation after infection by *Rhizobium* in *L. corniculatus* and red clover plants.[71–74] However, the use of transgenic plants produced from disarmed strains of *A. tumefaciens* would provide a better system for the expression of plant genes involved in nitrogen fixation, because the roots, unlike hairy roots, are phenotypically and physiologically normal. Such a system was used by Simons et al.[75] in order to study the regulation of leghemoglobin gene expression. Transgenic alfalfa plants were produced using *A. tumefaciens* carrying a chimeric gene consisting of a promoter from a gene coding for leghemoglobin fused to the coding region of chloramphenicol acetyltransferase (CAT), and high levels of CAT activity were found in the nodules but not in other parts of the plant.

Conclusions

Despite much progress, *A. tumefaciens*-mediated transformation of legumes is still not sufficiently effective in many species to allow for commercial use.

Further work is necessary on this and other successful means of legume transformation (e.g., biolistics in soybeans[59,76]) to allow the potential to be met for both the improvement of crops and the investigation of the basic biology of this important plant family.

References

1. **Binns, A. N. and Thomashow, M. F.,** Cell biology of *Agrobacterium* and transformation of plants, *Annu. Rev. Microbiol.,* 42, 575, 1988.

2. **Zambryski, P., Tempe, J., and Schell, J.,** Transfer and function of T-DNA genes from *Agrobacterium* Ti and Ri plasmids in plants, *Cell,* 56, 193, 1989.

3. **Klee, H., Horsch, R., and Rogers, S.,** *Agrobacterium*-mediated plant transformation and its further applications to plant biology, *Annu. Rev. Plant Physiol.,* 38, 467, 1987.

4. **Gasser, C. S. and Fraley, R. T.,** Genetically engineering plants for crop improvement, *Science,* 244, 1293, 1989.

5. **Weising, K., Schell, J., and Kahl, G.,** Foreign genes in plants: transfer, structure, expression and applications, *Annu. Rev. Genet.,* 22, 421, 1988.

6. **Horsch, R. B., Fry, J. E., Hoffmann, N. L., Eichholtz, D., Rogers, S. G., and Fraley, R. T.,** A simple and general method for transferring genes into plants, *Science,* 227, 1229, 1985.

7. **Duke, J. A., Ed.,** *Handbook of Legumes of World Economic Importance,* Plenum Press, New York, 1981.

8. **Salunke, D. K. and Kadam, S. S., Eds.,** *Handbook of World Food Legumes: Nutritional Chemistry, Processing Technology, and Utilization,* Vol. 1, 2, and 3, CRC Press, Boca Raton FL, 1989.

9. **Bajaj, Y. P. S., Ed.,** *Biotechnology in Agriculture and Forestry, Vol. 10, Legumes and Oilseed Crops I,* Springer-Verlag, Berlin, 1990.

10. **Mroginski, L. A. and Kartha, K. K.,** Tissue culture of legumes for crop improvement, in *Plant Breeding Reviews,* Vol. 2, Janick, J., Ed., AVI Publishing, Westport, CT, 1984.

11. **Hood, E. E., Fraley, R. T., and Chilton, M.-D.,** Virulence of *Agrobacterium tumefaciens* strain A281 on legumes, *Plant Physiol.,* 83, 529, 1987.

12. **Owens, L. D. and Cress, D. E.,** Genotypic variability of soybean response to *Agrobacterium* strains harboring the Ti or Ri plasmids, *Plant Physiol.,* 77, 87, 1985.

13. **Hood, E. E., Chilton, W. S., Chilton, M.-D., and Fraley, R. T.,** T-DNA and opine synthetic loci in tumors incited by *Agrobacterium tumefaciens* A281 on soybean and alfalfa plants, *J. Bacteriol.,* 168, 1283, 1986.

14. **Pedersen, H. C., Christiansen, J., and Wyndaele, R.,** Induction and *in vitro* culture of soybean crown gall tumors, *Plant Cell Rep.,* 2, 201, 1983.

15. **Byrne, M. C., McDonnell, R. E., Wright, M. S., and Carnes, M. G.,** Strain and cultivar specificity in the *Agrobacterium*-soybean interaction, *Plant Cell Tissue Organ Cult.,* 8, 3, 1987.

16. **Puonti-Kaerlas, J., Stabel, P., and Eriksson, T.,** Transformation of pea (*Pisum sativum* L.) by *Agrobacterium tumefaciens, Plant Cell Rep.,* 8, 321, 1989.

17. **Hawes, M. C., Robbs, S. L., and Pueppke, S. G.,** Use of a root tumorigenesis assay to detect genotypic variation in susceptibility of thirty-four cultivars of *Pisum sativum* to crown gall, *Plant Physiol.,* 90, 180, 1989.

18. **Hobbs, S. L. A., Jackson, J. A., and Mahon, J. D.,** Specificity of strain and genotype in the susceptibility of pea to *Agrobacterium tumefaciens, Plant Cell Rep.,* 8, 274, 1989.

19. **El Khalifa, M. D. and Lippincott, J. A.,** The influence of plant-growth factors on the initiation and growth of crown-gall tumours on primary pinto bean leaves, *J. Exp. Bot.,* 19, 61, 749, 1968.

20. **McClean, P., Chee, P., Held, B., Simental, J., Drong, R. F., and Slightom, J.,** Susceptibility of dry bean (*Phaseolus vulgaris* L.) to *Agrobacterium* infection: transformation of cotyledonary and hypocotyl tissues, *Plant Cell Tissue Organ Cult.,* 24, 131, 1991.

21. **Warkentin, T. D. and McHughen, A.,** Crown gall transformation of lentil (*Lens culinaris* Medik.) with virulent strains of *Agrobacterium tumefaciens, Plant Cell Rep.,* 10, 489, 1991.

22. **Gill, R., Eapen, S., and Rao, P. S.,** Transformation of the grain legume *Vigna acoitifolia* Jacq Marechal by *Agrobacterium tumefaciens,* regeneration of shoots, *Proc. Indian Acad. Sci. (Plant Sci.),* 98, 495, 1988.

23. **Webb, K. J.,** Transformation of forage legumes using *Agrobacterium tumefaciens, Theor. Appl. Genet.,* 72, 53, 1986.

24. **Armstead, I. P. and Webb, K. J.,** Effect of age and type of tissue on genetic transformation of *Lotus corniculatus* by *Agrobacterium tumefaciens, Plant Cell Tissue Organ Cult.,* 9, 95, 1987.

25. **Mariotti, D., Davey, M. R., Draper, J., Freeman, J. P., and Cocking, E. C.,** Crown gall tumorigenesis in the forage legume *Medicago sativa* L., *Plant Cell Physiol.,* 25, 473, 1984.

26. **Bevan, M. W., Flavell, R. B., and Chilton, M.-D.,** A chimaeric antibiotic resistance gene as a selectable marker for plant cell transformation, *Nature,* 304, 184, 1983.

27. **Lloyd, A. M., Barnason, A. R., Rogers, S. G., Byrne, M. C., Fraley, R. T., and Horsch, R. B.,** Transformation of *Arabidopsis thaliana* with *Agrobacterium tumefaciens, Science,* 234, 464, 1986.

28. **Shah, D. M., Horsch, R. B., Klee, H. J., Kishore, G. M., Winter, J. A., Tumer, N. E., Hironaka, C. M., Sanders, P. R., Gasser, C. S., Aykent, S., Siegel, N. R., Rogers, S. G., and Fraley, R. T.,** Engineering herbicide tolerance in transgenic plants, *Science,* 233, 478, 1986.

29. **de Block, M., Botterman, J., Vandewiele, M., Dockx, J., Thoen, C., Gossele, V., Movva, N. R., Thompson, C., Van Montagu, M., and Leemans, J.,** Engineering herbicide resistance in plants by expression of a detoxifying enzyme, *EMBO J.,* 6, 2513, 1987.

30. **Depicker, A., Stachel, S., Dhaese, P., Zambryski, P., and Goodman, H. M.,** Nopaline synthase: transcript mapping and DNA sequence, *J. Mol. Appl. Genet.,* 1, 561, 1982.

31. **Jefferson, R. A., Kavanagh, T. A., and Bevan, M. W.,** GUS fusions: β-glucuronidase as a sensitive and versatile gene fusion marker in higher plants, *EMBO J.,* 6, 3901, 1987.

32. **Fillatti, J. J., Kiser, J., Rose, R., and Comai, L.,** Efficient transfer of a glyphosate tolerance gene into tobacco using a binary *Agrobacterium tumefaciens* vector, *Bio/Technology,* 5, 726, 1987.

33. **Chabaud, M., Passiatore, J. E., Cannon, F., and Buchanan-Wollaston, V.,** Parameters affecting the frequency of kanamycin resistant alfalfa obtained by *Agrobacterium tumefaciens* mediated transformation, *Plant Cell Rep.,* 7, 512, 1988.

34. **Hinchee, M. A. W., Connor-Ward, D. V., Newell, C. A., McDonnell, R. E., Sato, S. J., Gasser, C. S., Fischoff, D. A., Re, D. B., Fraley, R. T., and Horsch, R. B.,** Production of transgenic soybean plants using *Agrobacterium*-mediated DNA transfer, *Bio/Technology,* 6, 915, 1988.

35. **Lulsdorf, M. M., Rempel, H., Jackson, J. A., Baliski, D. S., and Hobbs, S. L. A.,** Optimizing the production of transformed pea (*Pisum sativum* L.) callus using disarmed *Agrobacterium tumefaciens* strains, *Plant Cell Rep.,* 9, 479, 1991.

36. **White, D. W. R. and Greenwood, D.,** Transformation of the forage legume *Trifolium repens* L. using binary *Agrobacterium* vectors, *Plant Mol. Biol.,* 8, 461, 1987.

37. **De Kathen, A. and Jacobsen, H.-J.,** *Agrobacterium tumefaciens*-mediated transformation of *Pisum sativum* L. using binary and cointegrate vectors, *Plant Cell Rep.*, 9, 276, 1990.

38. **Godwin, I., Todd, G., Ford-Lloyd, B., and Newbury, H. J.,** The effects of acetosyringone and pH on *Agrobacterium*-mediated transformation vary according to plant species, *Plant Cell Rep.*, 9, 671, 1991.

39. **Shahin, E. A., Spielmann, A., Sukhapinda, K., Simpson, R. B., and Yashar, M.,** Transformation of cultivated alfalfa using disarmed *Agrobacterium tumefaciens*, *Crop Sci.*, 26, 1235, 1986.

40. **Garcia, J. A., Hile, J., and Goldbach, R.,** Transformation of cowpea *Vigna unguiculata* cells with an antibiotic resistance gene using a Ti-plasmid-derived vector, *Plant Sci.*, 44, 37, 1986.

41. **Baldes, R., Moos, M., and Geider, K.,** Transformation of soybean protoplasts from permanent suspension cultures by cocultivation with cells of *Agrobacterium tumefaciens*, *Plant Mol. Biol.*, 9, 135, 1987.

42. **Manners, J. M.,** Transformation of *Stylosanthes* spp. using *Agrobacterium tumefaciens*, *Plant Cell Rep.*, 6, 204, 1987.

43. **D'Halluin, K., Botterman, J., and De Greef, W.,** Engineering of herbicide-resistant alfalfa and evaluation under field conditions, *Crop Sci.*, 30, 866, 1990.

44. **Kuchuk, N., Komarnitski, I., Shakhovsky, A., and Gleba, Y.,** Genetic transformation of *Medicago* species by *Agrobacterium tumefaciens* and electroporation of protoplasts, *Plant Cell Rep.*, 8, 660, 1990.

45. **Hill, K. K., Jarvis-Eagan, N., Halk, E. L., Krahn, K. J., Liao, L. W., Mathewson, R. S., Merlo, D. J., Nelson, S. E., Rashka, K. E., and Loesch-Fries, L. S.,** The development of virus-resistant alfalfa, *Medicago sativa* L., *Bio/Technology*, 9, 373, 1991.

46. **Pezzotti, M., Pupilli, F., Damiani, F., and Arcioni, S.,** Transformation of *Medicago sativa* L. using a Ti plasmid derived vector, *Plant Breed.*, 106, 39, 1991.

47. **Deak, M., Kiss, G. B., Koncz, C., and Dudits, D.,** Transformation of *Medicago* by *Agrobacterium* mediated gene transfer, *Plant Cell Rep.*, 5, 97, 1986.

48. **Penza, R., Lurquin, P. F., and Filippone, E.,** Gene transfer of mature embryos with *Agrobacterium tumefaciens:* application to cowpea (*Vigna unguiculata* Walp), *J. Plant Physiol.*, 138, 39, 1991.

49. **Eapen, S., Köhler, F., Gerdemann, M., and Schieder, O.,** Cultivar dependence of transformation rates in moth bean after co-cultivation of protoplasts with *Agrobacterium tumefaciens*, *Theor. Appl. Genet.*, 75, 207, 1987.

50. **Manners, J. M.,** Transgenic plants of the tropical pasture legume *Stylosanthes humilis, Plant Sci.,* 55, 61, 1988.

51. **Pickardt, T., Meixner, M., Schade, V., and Schieder, O.,** Transformation of *Vicia narbonensis* via *Agrobacterium*-mediated gene transfer, *Plant Cell Rep.,* 9, 535, 1991.

52. **Mariotti, D., Fontana, G. S., and Santini, L.,** Genetic transformation of grain legumes: *Phaseolus vulgaris* L. and *P. coccineus* L., *J. Genet. Breed.,* 43, 77, 1989.

53. **Chee, P. P., Fober, K. A., and Slightom, J. L.,** Transformation of soybean *(Glycine max)* by infecting germinating seeds with *Agrobacterium tumefaciens, Plant Physiol.,* 91, 1212, 1989.

54. **Parrott, W. A., Hoffman, L. M., Hildebrand, D. F., Williams, E. G., and Collins, G. B.,** Recovery of primary transformants of soybean, *Plant Cell Rep.,* 7, 615, 1989.

55. **Zhou, J. H. and Atherly, A. G.,** *In situ* detection of transposition of the maize controlling element *(Ac)* in transgenic soybean tissues, *Plant Cell Rep.,* 8, 542, 1990.

56. **Puonti-Kaerlas, J., Eriksson, T., and Engström, P.,** Production of transgenic pea (*Pisum sativum* L.) plants by *Agrobacterium tumefaciens*-mediated gene transfer, *Theor. Appl. Genet.,* 80, 246, 1990.

57. **Wier, A. T., Thro, A. M., Flores, H. E., and Janyes, J. M.,** Transformation of *Lotononis bainesii* Baker with a synthetic protein gene using the leaf disk transformation-regeneration method, *Phyton,* 48, 123, 1988.

58. **Jackson, J. A. and Hobbs, S. L. A.,** Rapid multiple shoot production from cotyledonary node explants of pea (*Pisum sativum* L.), *In Vitro Cell. Dev. Biol.,* 26, 835, 1990.

59. **Finer, J. J. and McMullen, M.,** Transformation of soybean via particle bombardment of embryogenic suspension culture tissue, *In Vitro Cell. Dev. Biol.,* 27P, 175, 1991.

60. **Van Lijsebettens, M., den Boer, B., Hernalsteens, J.-P., and Van Montagu, M.,** Insertional mutagenesis in *Arabidopsis thaliana, Plant Sci.,* 80, 27, 1991.

61. **Fristensky, B., Horovitz, D., and Hadwiger, L. A.,** cDNA sequences for pea disease resistance response genes, *Plant Mol. Biol.,* 11, 713, 1988.

62. **Higgins, T. J. V., Newbigin, E. J., Spencer, D., Llewellyn, D. J., and Craig, S.,** The sequence of a pea vicilin gene and its expression in transgenic tobacco plants, *Plant Mol. Biol.,* 11, 683, 1988.

63. **Rerie, W. G., Whitecross, M., and Higgins, T. J. V.,** Developmental and environmental regulation of pea legumin genes in transgenic tobacco, *Mol. Gen. Genet.,* 225, 148, 1991.

64. **Altenbach, S. B., Kuo, C.-C., Staraci, L. C., Pearson, K. W., Wainwright, C., Georgescu, A., and Townsend, J.,** Accumulation of a Brazil nut albumin in seeds of transgenic canola results in enhanced levels of seed protein methionine, *Plant Mol. Biol.,* 18, 235, 1992.

65. **Webb, K. J., Woodcock, S., and Chamberlain, D. A.,** Plant regeneration from protoplasts of *Trifolium repens* and *Lotus corniculatus, Plant Breed.,* 98, 111, 1987.

66. **Hiatt, W. R., Kramer, M., and Sheehy, R. E.,** The application of antisense RNA technology to plants, in *Genetic Engineering Principles and Methods,* Vol. 11, Setlow, J. K., Ed., Plenum Press, New York, 1989, 49.

67. **Osborn, T. C., Alexander, D. C., Sun, S. S. M., Cardona, C., and Bliss, F. A.,** Insecticidal activity and lectin homology of arcelin seed protein, *Science,* 240, 207, 1988.

68. **Murdock, L. L., Huesing, J. E., Nielsen, S. S., Pratt, R. C., and Shade, R. E.,** Biological effects of plant lectins on the cowpea weevil, *Phytochemistry,* 29, 85, 1990.

69. **Altabella, T. and Chrispeels, M. J.,** Tobacco plants transformed with the bean *αai* gene express an inhibitor of insect α-amylase in their seeds, *Plant Physiol.,* 93, 805, 1990.

70. **Hilder, V. A., Gatehouse, A. M. R., Sheerman, S. E., Barker, R. F., and Boulter, B.,** A novel mechanism of insect resistance engineered into tobacco, *Nature,* 330, 160, 1987.

71. **Bogusz, D., Llewellyn, D. J., Craig, S., Dennis, E. S., Appleby, C. A., and Peacock, W. J.,** Nonlegume hemoglobin genes retain organ-specific expression in heterologous transgenic plants, *Plant Cell,* 2, 633, 1990.

72. **Szabados, L., Ratet, P., Grunenberg, B., and de Bruijn, F. J.,** Functional analysis of the *Sesbania rostrata* leghemoglobin *glb3* gene 5′-upstream region in transgenic *Lotus corniculatus* and *Nicotiana tabacum* plants, *Plant Cell,* 2, 973, 1990.

73. **Miao, G.-H., Hirel, B., Marsolier, M. C., Ridge, R. W., and Verma, D. P. S.,** Ammonia-regulated, expression of a soybean gene encoding cytosolic glutamine synthetase in transgenic *Lotus corniculatus, Plant Cell,* 3, 11, 1991.

74. **Beach, K. H. and Gresshoff, P. M.,** Characterization and culture of *Agrobacterium rhizogenes* transformed roots of forage legumes, *Plant Sci.,* 57, 73, 1988.

75. **Simons, A., Schreier, P. H., Schell, J., and de Bruijn, F. J.,** *Agrobacterium* mediated transformation of alfalfa *(Medicago sativa):* use of transgenic plants carrying chimeric *lb-cat* genes to study nodule-specific *lb* gene induction, in *Nitrogen Fixation: Hundred Years After,* Bothe, H., de Bruijn, F. J., and Newton, W. E., Eds., Gustav Fischer, Stuttgart, 1988, 643.

76. **Christou, P., McCabe, D. E., Martinell, B. J., and Swain, W. F.,** Soybean genetic engineering — commercial production of transgenic plants, *Trends Biotechnol.,* 8, 145, 1990.

77. **Jordan, M. C. and Hobbs, S. L. A.,** Evaluation of a cotyledonary node regeneration system for *Agrobacterium*-mediated transformation of pea (*Pisum satiuum* L.), *In Vitro Cell. Dev. Biol.,* 29P, 77, 1993.

78. **Jordan, M. C. and Hobbs, S. L. A.,** Unpublished results.

Biotechnological Applications of Haploids

A. M. R. Ferrie,[1] C. E. Palmer,[2] and W. A. Keller[1]

[1]*Plant Biotechnology, National Research Council of Canada, Saskatoon, Saskatchewan, Canada*

[2]*Department of Plant Science, University of Manitoba, Winnipeg, Manitoba, Canada*

Introduction

Although the occurrence of haploidy in Angiosperms was reported several decades ago,[1] it was the report of Guha and Maheshwari[2] which stimulated widespread interest in the *in vitro* production of haploids through the culture of gametophytic cells and tissues. They demonstrated the regeneration of haploid plants from cultured anthers of *Datura innoxia* Mill. The totipotency of microspores was, therefore, recognized and later studies indicated that isolated microspores of tobacco, *Nicotiana tabacum* L., also regenerated haploid plants in culture.[3] Since spontaneous occurrence of haploids is rare, the importance of tissue culture for the large scale production of these plants was quickly recognized. The value of these plants to plant breeding and genetic studies was evident, and efforts were concentrated on the development of haploids from economically important plant species. Since this early report, haploid production has been shown in well over 200 species.[4–6] Table 1 lists some species in which haploids have been produced since 1985. For a listing of other species the reader is referred to reviews by Maheshwari et al.[6] and Dunwell.[48]

0-8493-8262-9/94/$0.00+$.50

Table 1 *Haploid and diploid plant production since 1985*

Species	Ref.	Species	Ref.
Aconitum carmichaeli	7	*Morus*	29
Aesculus carnea	8	*Nicotiana otophora*	30
Allium cepa	9	*Nicotiana plumbaginifolia*	30
Asparagus officinalis	10	*Nicotiana sylvestris*	31
Avena sativa	11	*Nicotiana tabacum*	32
Beta vulgaris	12	*Oryza sativa*	33
Brassica campestris	13	*Papaver somniferum*	34
Brassica carinata	14	*Peltophorum pterocarpum*	35
Brassica juncea	15	*Petunia* (ovary)	36
Brassica napus	16	*Populus maximowiczii*	37
Capsicum annuum	17	Poplars	38
Carica papaya	18	*Scillia indica*	39
Citrus aurantifolia	19	*Secale cereale*	40
Citrus	20	*Sinocalamus latiflora*	41
Digitalis obscura	21	*Solanum chacoense*	42
Gerbera jamesonii	22	*Solanum papita*	43
Helianthus annuus	23	*Solanum tubersum*	44
Hordeum vulgare	24	Sorghum	45
Hordeum spontaneum	25	*Triticale*	46
Iochroma warscewiczii	26	*Triticum aestivum*	24
Linum usitatissimum	27	*Zea mays*	47
Lolium perenne	28		

The purpose of this review is to examine the biotechnological applications of haploid cell cultures. Therefore, treatment of methods of haploid production and factors affecting such production will be brief, even though an understanding of the use of haploids requires some knowledge of their production. There are a number of excellent reviews on this topic to which the reader can refer.[6,48,49]

Methods of Haploid Production

Four major methods of haploid plant production are currently used: (1) culture of anthers and microspores, (2) culture of unfertilized ovules, (3) chromosome elimination (*bulbosum* method), and (4) those produced from wide crosses or through pollen or chemical treatment. These will be discussed briefly.

Androgenesis

The process of embryo development from microspores is referred to as androgenesis. In many species the culture of isolated microspores or anthers is the most frequently used method of haploid production. In this method, every microspore is potentially capable of regenerating an embryo, and each plant would therefore represent the variation which exists in the population of microspores. Plant regeneration occurs either by direct embryogenesis, as is the case in most dicots, or through callus formation followed by embryogenesis or organogenesis, which is common in cereals. Whether isolated microspores or anthers respond to culture depends on the species. In monocots, anther culture is the preferred method, while in some dicots, such as the *Brassicas,* isolated microspore culture is very successful. In *Brassica napus* L., isolated microspores were shown to be much more efficient in embryo production compared to anthers.[50] There are a number of factors affecting microspore embryogenesis, including genotype, donor plant growth conditions, stage of microspore development, composition of the culture medium, and environmental conditions during culture.[4,51] The frequency of embryo production will depend on whether or not these conditions are optimal. For additional information, the reader is referred to reviews by Heberle-Bors[52] and Maheshwari et al.[6]

Gynogenesis

The process of haploid development from the culture of unfertilized ovaries or ovules is referred to as gynogenesis. Under the appropriate culture conditions, cells of the embryo sac undergo division and organize an embryo without the benefit of fertilization. By this procedure, haploids have been produced in a number of species in several families.[53] The literature on this topic has been reviewed recently by Yang and Zhou.[53] Compared to androgenesis, the number of embryos produced by gynogenesis is low, and this process appears unsuitable for large-scale haploid production; however, it may be an alternative to microspore-derived haploids where defective pollen is produced or if these are unresponsive to culture. The quality of plants produced by gynogenesis may be better than those from androgenesis.[54]

Chromosome Elimination

With interspecific crosses, maternal haploids can be produced through elimination of the pollinator chromosomes. This method of haploid production has been applied to barley, *Hordeum vulgare,* where crosses with *Hordeum bulbosum* resulted in the recovery of *H. vulgare* haploids.[55] Following pollination and fertilization, embryo rescue is required to permit haploid embryo development, since endosperm degradation prevents *in planta* embryo development. This technique of haploid regeneration has been applied to other species such as *Triticum*.[56–59] In cereals, the *bulbosum* method can produce haploids efficiently, and barley varieties have been produced using this approach.[55]

Other Methods of Haploid Production

In addition to the above mentioned methods, haploids can be obtained by other procedures. In barley, haploid genes controlling the appearance of haploid embryos *in vivo* have been identified.[60] In some cases, such as in *Solanum tuberosum,* certain types of crosses yielded relatively high frequencies of haploids.[61,62] In some species, chemical treatment or pollination with inactivated pollen resulted in haploid recovery.[63,64] One potentially promising procedure for haploid production in wheat is the wheat × maize cross[65] in which high frequencies of haploids were recovered.[66,67] This procedure also involves embryo rescue and resembles the *bulbosum* method. Reports indicate that this method of haploid production in wheat is superior to production by the *bulbosum* method.[66,68–70]

Factors Affecting Haploid Production

The frequency of embryogenesis depends on a number of factors. Most of the studies conducted have concentrated on the factors influencing embryogenesis from microspores/anthers, and limited information is available on the factors influencing embryogenesis from ovaries/ovules.[53] The discussion below is primarily focused on haploids derived from microspore/anther culture. For discussion of factors relevant to haploidy by the *bulbosum* and wheat × maize crosses, refer to Sitch and Snape[57] and Laurie and Bennett.[65]

Plant

Genotype — Genotype plays a major role in influencing the embryogenic response of anthers/microspores and ovules. Genotype screening studies have resulted in different embryogenic response for different genotypes.[43,71,72] Studies have indicated that microspore embryogenesis is conditioned by nuclear genes. In *Solanum tuberosum,* more than one gene is involved, and these are recessive;[73] in *Triticum aestivum,* the most responsive genotypes are those possessing the 1B/1R translocation.[74,75]

Donor plant growing conditions — The environmental conditions in which donor plants are grown can influence embryogenesis. These conditions include temperature,[76] photoperiod,[76,77] light intensity,[78,79] time of year,[78] and nutrient supply.[80] A reduced growing temperature has been found to be beneficial in a number of species, e.g., *Brassica*.[4,81,82]

Developmental stage of the pollen — The developmental stage which is most responsive in terms of embryogenesis varies with species; however, this is usually between the early uninucleate to the early binucleate stages. Older and younger microspores may cause the development of toxic compounds in culture which could result in a decrease in embryogenesis[83] and abnormal, stunted embryos.[84] The technique (i.e., anther culture or microspore culture) used also can influence the optimum developmental stage.[85–87]

Culture Methods

Pretreatments — Embryogenesis has been enhanced when buds, spikes, florets, or anthers have been subjected to certain pretreatments. These include chemical application to donor plants,[88] low temperature,[89–93] treatment with colchicine,[94] reduced atmospheric pressure,[95] osmotic shock,[96] gamma irradiation, and ethanol stress.[97]

Media composition — The composition of the culture medium is a critical factor influencing embryogenesis. The factors which have been shown to influence embryogenesis are carbon source and concentration,[98,99] nitrogen source and concentration,[100,101] growth regulator type and composition,[102–104] ethylene inhibitors,[44,105] ethylene promotors,[106,107] and activated charcoal.[43,108] Physical factors of the media also can influence embryogenesis. These include liquid vs. solid media,[108,109] type and concentration of gelling agents,[110,111] and anther orientation.[112]

Utilization

With the production of haploids from a number of species, the initial interest to plant breeders was in the enormous value of the doubled haploids for the production of homozygous lines in a single step. Another feature of the microspore-derived plants is that they offer a broad spectrum from which to select. In addition, there are a number of other uses including mutagenesis, selection of gametoclonal variants, and in studies of embryogenesis. The purpose of this section is to review the current uses of haploid cell cultures and to highlight the advantages compared to other tissue culture techniques.

Plant Breeding and Genetics

One obvious advantage of haploid production to plant breeding is the ability to rapidly produce homozygous lines without the need for exhaustive inbreeding. It generally takes about 10 years to develop a variety through a conventional plant breeding program, whereas the use of haploids and double haploids can reduce this period by 3 to 4 years.[113] Once haploid embryos or plantlets are produced, doubled haploids are usually recovered by methods of chromosome doubling.[114] In addition, a significant population of spontaneously doubled haploids can occur;[115] however, this depends on species and genotype. The use of this technique to expedite a breeding program is evident and has been

Table 2 *Doubled haploid varieties*

Species	Variety	Ref.
Hordeum vulgare	Mingo	123
	Rodeo	124
	Gwylan	125
Nicotiana tabacum	F 211	126
	Tan-Yuh no 1	127
	NC 744	128
	LMAFC 34	129
Oryza sativa	Hua Yu 1	130
	Hua Yu 2	130
Triticum aestivum	Jinghua No 1	131
	Florin	132

utilized in the improvement of a number of crops, including tobacco, potato, oilseed-rape, rice, rye, barley, wheat, and maize.[116–122] In some cases, cultivars have been released which originated from doubled haploids (Table 2).[133,134]

Doubled haploids are particularly useful in studies of recessive traits since there are no heterozygous genotypes. Genetic ratios are simpler and fewer plants must be screened to find a particular genotype.[135,136] This technique was exploited in the development of *Brassica napus* plants having several recessive traits and in studies of erucic acid inheritance.[137,138] The use in analysis of inheritance, linkages, and quantitative traits has been reported.[139]

Double haploids have an additional advantage, i.e., the culture of anthers or microspores to plantlets can potentially result in a range of gamete recombinations which may not be achieved by conventional means.[134] Therefore, the potential exists for introducing alien genes into cultivated species as was reported for wheat.[140–142] This is possible because, especially in the case of F_1 hybrids, the gametes from which plants are derived represent an array of chromosomes and gene combinations of the hybrid. This aspect has been thoroughly discussed by Morrison and Evans.[5] With haploids, nonhomologous chromosomal exchanges can occur during mitosis leading to the appearance of novel recombinants. This may be a valuable source of genetic variation.[143] Haploids can be used to introduce wild germplasm into cultivated crops as was demonstrated in *Solanum tuberosum* L.[144] The advantage outlined by Jansky et al.[144] is that haploids originating from wild species can be evaluated for desirable characteristics at the haploid level.

Even with these important features of doubled haploids, their full exploitation in a breeding program requires that there not be significant differences between these populations and those derived from conventional techniques. For some characteristics, such as fatty acid composition in *Brassica napus,* doubled haploid populations are similar to those derived by single seed descent.[145,146] In other cases, differences are evident, leading to the suggestion of *in vitro* selection pressure and gamete selection during culture.[147–149] Such differences are classified as gametoclonal variation, and their usefulness will be addressed elsewhere in this chapter.

Another feature of haploidy (dihaploid) was illustrated in *Solanum tuberosum* L., for which dihaploid × dihaploid or dihaploid × tetraploid crosses were always superior in yield compared to parental lines.[143] Such heterosis has been demonstrated in *Brassica napus* L. where hybrid vigor for plant height resulted from crosses between androgenetically derived plants.[122]

Mutation and Selection

Gametic haploids or somatic cells derived from haploid plants make ideal materials for mutant selection. By the very nature of haploids, all recessive and dominant traits are readily expressed and are easily selectable in culture. Such traits can be fixed by chromosome doubling to achieve homozygosity. This is unlike diploid cell selection where recessives can be hidden in the heterozygote. Since selection can be made at the haploid or doubled haploid state, undesirable traits can be readily discarded instead of being carried for generations in the heterozygote.

Haploid tissues of *Nicotiana tabacum* L. and *Brassica napus* have been employed in mutant selection involving selection pressure imposed by pathotoxins or their analogues.[150,151] Several herbicide-resistant plants of tobacco have been selected using haploid cell cultures.[152,153] In addition, many auxotrophs have been selected from such cultures.[154,155] In some cases, amino acid analogues were employed in the selection of resistant cells which exhibited increased accumulation of the corresponding amino acid in regenerated plants.[156,157] A combination of mutagenesis and selection on a medium with 5-methyltryptophan resulted in isolation of tyrosine and phenylalanine overproducing dihaploid potato mutants.[158]

Chemical and physical mutagens are frequently employed in some of these selection procedures. In *Nicotiana plumbaginifolia,* amino acid mutants were selected following UV mutagenesis of haploid protoplasts.[159] The use of 5-bromodeoxyoridine (BUdR) enrichment resulted in selection of a number of mutants, and fertile plants were recovered following complementation through protoplast fusion. With *Hyoscyamus miticus* cells, the chemical mutagen N-methyl-N^1-nitro-N-nitroso guanidine (MNNG) allowed recovery of histidine auxotrophs.[160,161] Nitrate reductase mutants of *Petunia hybrida* were selected following X-ray mutagenesis and cell culture on a medium containing chlorate.[162] A similar procedure was used to obtain temperature-sensitive tobacco mutants from haploid protoplasts.[163]

Although the use of haploid cell cultures in mutation is widely applicable, whole plant regeneration from selected cells may not be easily achieved. Therefore, mutagenesis of cells, such as microspores or haploid embryos which have a high regenerative potential, is desirable. Plants resistant to sulfonylurea and imidazolinone herbicides were selected in *Brassica napus* cultures by mutagenizing microspores with ethylnitrosourea (ENU).[164,165] In the sulfonylurea-resistant mutant, this trait was semidominant and involved alteration in acetohydroxy acid synthase, the enzyme which is inhibited in susceptible plants. Microspores or microspore-derived haploids are also useful in the selection of mutants for storage product accumulation. Mutants with high oleic acid were isolated from mutagen-treated microspores of *B. napus*,[166] and other mutants have been reported.[167,168]

Apart from the high regenerative potential of microspores, there are other advantages for their use in mutagenesis. A large number of uniform cells can be exposed to chemical or physical mutagens in a relatively small space and mutants can be selected by the appropriate selection pressure. In the case of herbicide-tolerant mutants, the chemical can be incorporated directly into the culture medium. Where regeneration occurs by embryogenesis, the cotyledon can be analyzed for storage product accumulation and plants recovered from the rest of the embryo. In this way, rapid and reliable screening can be done.[168] Although such advantages are evident, there are surprisingly few other reports of the successful use of microspores in mutagenesis.[169,170] One potential limitation is the limited number of species in which microspore embryogenesis can be reliably achieved. There may be negative effects of mutagens on the regeneration capacity of the microspores. However, it has been reported that low levels of irradiation, ethanol stress, and low concentrations of chemical mutagens enhance pollen embryogenesis.[97,171–173] Successful use of mutagens requires optimization of conditions to ensure maximum retention of regeneration capacity. The timing of mutagen application may be critical as high frequency embryogenesis requires an undisturbed incubation period during the early stages of culture.[168] Mutants have been recovered following mutagenesis of pollen mother cells,[174] freshly isolated microspores,[170] and microspores one day after isolation.[167] However, these are not general procedures and efficient use of induced mutations in this system will require establishment of a reliable protocol.

Gametoclonal Selection

Variations similar to somaclonal variation[175] have been recorded among gamete-derived plants; these variations are defined as gametoclonal variations.[5,176] This topic has been thoroughly discussed in several reviews.[5,176] Morrison and Evans[5] have outlined the basic differences between these two processes as (1) the ready expression of any recessive trait in the haploid state or following chromosome doubling. This is in contrast to somaclones where selfing is required to detect recessives. The other difference (2) is that gametoclones require chromosome doubling for fertility and the process can itself lead to variation in the resulting plants.[177,178] An additional feature noted by Morrison and Evans[5] is the possibility of heterozygosity in the doubled haploids if genetic variation occurs after chromosome doubling.

The basis for gametoclonal variation is largely unexplained, but there are a number of possibilities such as mutations, genetic recombination, residual heterozygosity, unreduced gametes, organelle genome modification, and whether the vegetative or generative cell is the origin of haploid embryos. These are discussed to some extent by Morrison and Evans.[176] In anther- and microspore-derived plants, chromosomal alterations have been observed.[134,179] DNA amplification and other changes were observed in anther-derived doubled haploids of tobacco,[180,181] as well as instances of aneuploidy in pollen-derived rice plants.[182] These are likely causes of observed gametoclonal variation. The culture conditions may contribute to some of these variations although specific components have not been identified.[183,184] In this regard, all reports indicate much less variation among gynogenetic haploids compared to androgenetic haploids.[183,184] This may reflect the difference in culture conditions of the two systems.

As is the case with somaclonal variation, gametoclonal variants may be useful sources of variation for crop improvement. These variations have been employed in selection for disease resistance in a number of crop species.[150,185–192] When the pathogen produces a toxin, the toxin can be used to screen haploid cells before plant regeneration.[150,187–189] In other cases, doubled haploid plants were screened for disease resistance.[185,186,190,191] In addition to disease resistance, other useful variations have been reported. Doubled haploids exhibiting genetic variation for seed weight, leaf size, plant height, and increased protein content were isolated from rice anther culture.[125,193] Heritable variations for yield, morphology, and alkaloid content were found among gametoclones of tobacco.[194,195] Given this expression of variability, it is evident that the system lends itself to a variety of selection strategies, and with the appropriate selective agents, variants for a number of stress factors could be isolated. The theoretical aspects of microspore screening for development of stress tolerant genotypes has been discussed.[196]

Gene Transfer

The haploid system is well-suited for genetic transformation. This is due to the ease of handling a large number of uniform cells or embryos, usually of high regenerative potential, and the ability to duplicate the introduced trait during chromosome doubling for doubled haploid production. There are a number of techniques generally used for the introduction of foreign genes into plant cells, most of which have been used to some extent with haploid cells or tissues.

Transgenic *Brassica napus* plants were reported from microinjection of DNA into surface cells of microspore-derived embryos.[197] Transformation efficiencies were 27 to 51%, although the primary regenerants were chimeric. The most frequently used technique is explant cocultivation with *Agrobacterium*, and transformants for herbicide tolerance and antibiotic resistance have been reported.[168,198,199] In these reports, whole embryos or embryo segments were incubated with *Agrobacterium* and transformed plants regenerated by secondary embryogenesis. Transformation frequencies of 0.8 to 3.0% were observed.

Successful recovery of transformed plants following electroporation or the application of polyethylene glycol (PEG) to gametic tissues has not been achieved; however, these methods have been used in the delivery of DNA into microspores.[200,201] Electroporation may prove to be a particularly effective technique for DNA delivery into microspores or embryos, especially if immature, thin-walled microspores are used. In immature zygotic embryos of maize, a combination of mild enzymatic treatment and electroporation allowed efficient recovery of transformed plants.[202] Haploid embryos are well-suited to this technique which should increase the efficiency of transformation. Other techniques, such as DNA uptake by imbibition, were successful in seed-derived embryos and could easily be extended to haploid embryos.[203] Any gene transfer method used on haploid cells should not seriously diminish the regeneration capacity of these cells. Transformation studies with embryos and microspores indicate a slight reduction in cell viability with some DNA delivery methods.[200,201,204]

Protoplast Fusion

Haploid protoplasts are useful in studies of mutation and selection as described in earlier sections of this review. There are many instances of successful plant regeneration from haploid protoplasts, especially in cereals where embryogenic suspension cultures are established from haploid tissues.[205,206] Besides protoplast culture for mutant selection, there are other aspects of this technology which are potentially useful. Fusion of gametic protoplasts can overcome incompatibility barriers. Also, in this approach the recovery of diploid or tetraploid units is more likely compared to normal somatic fusions. Chromosome elimination, associated with somatic hybridization, may be reduced. Haploid cell fusion was used to recover diploid *Brassica napus* plants having cytoplasmic male sterility and herbicide tolerance.[207,208] For additional information on protoplast isolation and fusion from haploid tissues, the reader is referred to Bajaj.[209] Another aspect of haploid cell fusion is the production of triploids via gameto-somatic fusion. This has been recorded in *Nicotiana*,[210,211] and theoretical considerations relevant to the use of this technique in triploid production have been advanced.[212] A potential problem associated with this technique is the difficulty in recovering viable protoplasts from microspores capable of cell division. Methods are available for protoplast isolation from these cells, but sustained cell division is still a problem.[213–215] In *Allium* species, viable protoplasts were isolated from microspores of a number of genotypes following relatively short-term exposure to cell wall degrading enzymes.[215] If conditions for sustained cell division can be developed, gamete-gamete fusions and plant recovery may be possible. Since it is possible to select embryogenic microspores from a general population of microspores,[216,217] such enrichment before protoplast isolation may enhance subsequent culture and cell division.

Biochemical and Physiological Studies

The developmental sequence of microspore-derived haploid embryos closely approximates the zygotic counterpart, and many of the biochemical pathways leading to biosynthesis and accumulation of storage products are similar.[218] A comparison of zygotic and microspore-derived embryos of *Brassica napus* showed that during the early stages of development, the total fatty acids per unit weight was similar.[219,220] Fatty acid composition in the two types of embryos was also similar.[220,221] Haploid embryos, especially microspore-derived embryos, are therefore potentially useful in studies of basic biochemistry and physiology, not only of storage product accumulation, but of embryogenesis in general. Embryos can be obtained in large numbers from responding genotypes. Uniform developmental stages are readily available, and it is relatively easy to study the influence of growth regulators, media factors, and environmental influences on embryogenesis and metabolism. Zygotic embryos can be used in such studies but must first be excised at the appropriate stage and adapted to culture conditions which will allow an approximation of *in vivo* development.

An additional feature of microspore-derived embryos, at least in *Brassica napus,* is that they are rich sources of enzymes involved in lipid biosynthesis.[221] They are also useful for *in vitro* screening for oil quality.[222] In a survey of the literature, it is apparent that most of the biochemical and physiological studies using haploid embryos relate to *Brassica* species. These studies have been used to examine the influence of osmoticum and abscisic acid on lipid accumulation in *B. napus.*[223,224] Microspore-derived embryos have been used extensively in studies of lipid biosynthesis and as a model to study gene expression in zygotic embryos.[225–232] One area of considerable interest in lipid biochemistry is the biosynthesis and regulation of very long chain fatty acids such as erucic acid. Microspore-derived embryos have been used to define the enzyme systems involved in their biosynthesis.[233–235] Studies of storage protein synthesis and accumulation in microspore-derived embryos are less frequent, but the role of growth regulators and osmotic conditions in storage protein gene expression have been examined.[221,236,237] These embryos were also used in comparative studies of pyruvate-kinase isozymes of germinating seeds and developing embryos to determine the possible role of this enzyme in metabolite partitioning.[238]

The haploid embryo system and suspension culture cells derived from these embryos have been used for studies of chilling tolerance, freezing tolerance, chlorophyll metabolism, desiccation tolerance, and physiological aspects of embryo maturation.[226,239–247] The advantages outlined for haploid embryo usefulness in biochemical studies are equally applicable to their use in physiological studies. In *Brassica napus,* there are reports of the use of microspore-derived embryos in selection for freezing tolerance.[248,249] In these studies, plants derived from microspores exposed to low temperatures showed increased cold tolerance. Microspore-derived embryos have also been used in comparative studies of glucosinolate metabolism during embryo development.[250] Given these uses, it appears that the haploid embryo system can be a valuable model for studies of a variety of metabolic functions.

Germplasm Storage and Other Uses

The ability to store embryogenic microspores or embryos is of value in allowing long-term use for various studies and as stock materials from which plants can be recovered. In highly responsive genotypes, embryos are often generated in larger numbers than can be conveniently handled. In some species, such as woody perennials, microspores may be produced for a limited time during the growing season, and storage without deterioration would extend their use. Any storage method must not alter the regeneration capacity of the cells and should preserve their genetic integrity.

One method of storage is by cryopreservation at liquid nitrogen temperatures. This has been successfully applied to a number of species. The literature on cryopreservation of pollen-derived embryos has been reviewed by Bajaj.[251] Factors such as genotype, developmental stage of the embryos, water content, and the use of cryoprotectants are important in plant recovery following low temperature storage.[115,251–255] Isolated microspores of *Brassica napus* were cryopreserved without loss of embryogenic capacity. An added advantage of cryopreservation of microspores is the possible recovery of a large number of spontaneously diploid plants following embryogenesis.[115] This removes the necessity for chromosome doubling techniques, which may themselves result in abnormalities.

Haploid embryos can be desiccated to 15% moisture content and stored for an extended period.[245–247] The stage at which embryos are desiccated is important for successful plant recovery. The use of abscisic acid and suitable osmotica are essential to the embryo maturation process. Development of this procedure for embryo storage may be very useful as storage can be done at room temperature as opposed to cryopreservation. An additional benefit of this technique is the possible use of these embryos as artificial seeds. The microspore-derived embryos can be encapsulated in a protective coat and subsequently germinated as true seeds. The technique has been applied to somatic embryos derived from a number of species, including alfalfa.[256,257] In barley, microspore-derived embryos were encapsulated in calcium alginate beads. These artificial seeds maintained the ability to germinate for several months.[258] Because of their uniformity, microspore-derived embryos are well-suited for use as artificial seeds.[259]

Conclusions and Future Prospects

Haploids and doubled haploids have been produced from a number of species via microspore/anther culture, ovule/ovary culture, chromosome elimination, or special pollination mechanisms. There are a number of factors influencing embryogenesis and plant regeneration. The conditions required for embryogenesis differ depending on the genotype and species.

Haploids and doubled haploids are of benefit in plant breeding, mutagenesis, gene transfer, and biochemical and physiological studies.

The main advantage is the reduction in time to develop new varieties. A conventional plant breeding program takes about 10 years to develop a variety, whereas the use of microspore culture can reduce this length of time by 3 to 4 years. Second, one can rapidly fix traits in the homozygous condition. Homozygous, doubled-haploid lines can be produced in one generation, thereby eliminating 3 to 4 years of selfing or backcrossing to produce true-breeding lines. Third, the efficiency at which selection can take place is also improved. The phenotype of the plant is not masked by dominance effects. Traits conditioned by recessive genes can be easily identified. Fourth, a much smaller proportion of doubled haploids is required when screening for desirable recombinants than would be the case for conventional diploid populations.

Haploid cells, including microspores, may be quite useful for gene transfer prior to embryo induction. This would ensure recovery of nonchimeric transformants since each transformed cell regenerates a complete embryo. In addition, the usually high regeneration potential of haploid embryos, combined with proven means of gene transfer such as electroporation and biolistic techniques, should be exploited in transformation experiments.

There are very few cases of haploid protoplast fusion and recovery of hybrids, although this technique is of value in both diploid and triploid plant production. This may be due to difficulties in the isolation of protoplasts from gametes and chromosome instability in haploid cell cultures.

One of the emerging uses of haploid cultures, especially microspore-derived embryos, is in studies of basic physiology and biochemistry. This system provides a useful model for studies of embryogenesis and has been well-utilized in the *Brassicas* for analysis of lipid and protein biosynthesis. Development of conditions for efficient embryo production in other species should allow similar studies on carbohydrate biosynthesis in species where this is the main storage product.

The uniformity and abundance of haploid, microspore-derived embryos should allow for their use as artificial seeds. This is particularly important if chromosome doubling can be effected prior to embryo production. In this way, the resulting plantlets are already diploidized. Artificial seed technology has been developed for a few species using somatic embryos. Given the ease of embryo production from microspores in responding genotypes and the present state of knowledge of factors regulating embryo maturation and storage product accumulation, this should be a very useful system for artificial seed development.

The extensive use of haploids is limited because of the lack of reliable protocols for haploid embryo induction in some species (e.g., legumes and woody species) and because of the genotypic differences within a species. This is probably due to the absence of basic information on the biochemistry and physiology of embryogenesis. Progress in understanding the haploid embryo induction process will enhance their use both in basic plant biology and in crop improvement.

Utilization may be hampered by variation among haploid and doubled haploids and the quality of these plants. As an example, there is the occurrence of extensive albinism in haploids derived from cereal microspores which reduces the usefulness of these plants. Some culture-induced variation is potentially useful, and disease resistance and agronomic traits have been selected from haploid cultures. *Brassica napus* microspores subjected to low temperatures regenerated plants with increased cold tolerance.[249] It may, therefore, be possible to use other selection pressures (pathotoxins heavy metals, herbicides, or salt stress) to screen potentially embryogenic microspores before plant regeneration. In this way, a large number of uniform cells can be exposed to selection pressure in a relatively small space. A potential disadvantage is a possible reduction in embryogenesis as a result of these treatments.

References

1. **Blakeskee, A. F., Belling, J., Farnham, M. E., and Bergner, A. D.,** A haploid mutant in the Jimson weed, *Datura stramonium, Science,* 55, 646, 1922.

2. **Guha, S. and Maheshwari, S. C.,** *In vitro* production of embryos from anthers of *Datura, Nature,* 204, 497, 1964.

3. **Nitsch, J. P. and Nitsch, C.,** Haploid plants from pollen grains, *Science,* 163, 85, 1969.

4. **Keller, W. A., Arnison, P. G., and Cardy, B. J.,** Haploids from gametophytic cells — recent developments and future prospects, in *Plant Tissue and Cell Culture,* Green, C. E., Somers, D. A., Hackett, W. P., and Bresboer, D. D., Eds., Alan R. Liss, New York, 1987, 223.

5. **Morrison, R. A. and Evans, D. A.,** Haploid plants from tissue cultures: new plant varieties in a shortened time frame, *Biotechnology,* 6, 684, 1988.

6. **Maheshwari, S. C., Rashid, A., and Tyagi, A. K.,** Haploids from pollen grains — retrospects and prospects, *Am. J. Bot.,* 69, 865, 1982.

7. **Hatano, K., Shoyama, Y., and Nishioka, I.,** Somatic embryogenesis and plant regeneration from the anther of *Aconitum carmichaeli* Debx, *Plant Cell Rep.,* 6, 446, 1987.

8. **Radojevic, L., Djordjevic, N., and Tucic, B.,** *In vitro* induction of pollen embryos and plantlets in *Aesculus carnea* Hayne through anther culture, *Plant Cell Tissue Organ Cult.,* 17, 21, 1989.

9. **Keller, J.,** Culture of unpollinated ovules, ovaries, and flower buds in some species of the genus *Allium* and haploid induction via gynogenesis in onion (*Allium* cepa L.), *Euphytica,* 47, 241, 1990.

10. **Feng, X. R. and Wolyn, D. J.,** Development of haploid asparagus embryos from liquid cultures of anther-derived calli enhanced by ancymidol, *Plant Cell Rep.,* 12, 281, 1993.

11. **Rines, H. W. and Dahleen, L. S.,** Haploid oat plants produced by application of maize pollen to emasculate oat florets, *Crop Sci.,* 30, 1073, 1990.

12. **Bossoutrot, D. and Hosemans, D.,** Gynogenesis in *Beta vulgaris* L.: from *in vitro* culture of unpollinated ovules to the production of doubled haploid plants in soil, *Plant Cell Rep.,* 4, 300, 1985.

13. **Baillie, A. M. R., Epp. D. J., Hutcheson, D., and Keller, W. A.,** *In vitro* culture of isolated microspores and regeneration of plants in *Brassica campestris, Plant Cell Rep.,* 11, 234, 1992.

14. **Arora, R. and Bhojwani, S. S.,** Production of androgenic plants through pollen embryogenesis in anther cultures of *Brassica carinata* A. Braun, *Biologia Plant.,* 30, 25, 1988.

15. **Sharma, K. K. and Bhojwani, S. S.,** Microspore embryogenesis in anther cultures of two Indian cultivars of *Brassica juncea* (L.) Czern., *Plant Cell Tissue Organ Cult.,* 4, 234, 1985.

16. **Huang, B., Bird, S., Kemble, R., Simmonds, D., Keller, W., and Miki, B.,** Effects of culture density, conditioned medium and feeder cultures on microspore embryogenesis in *Brassica napus* L. cv. Topas, *Plant Cell Rep.,* 8, 594, 1990.

17. **Morrison, R. A., Koning, R. E., and Evans, D. A.,** Anther culture of an interspecific hybrid of *Capsicum, J. Plant Physiol.,* 126, 1, 1986.

18. **Tsay, H. S. and Su, C. Y.,** Anther culture of papaya (*Carica papaya* L.), *Plant Cell Rep.,* 4, 28, 1985.

19. **Chaturvedi, H. C. and Sharma, A. K.,** Androgenesis in *Citrus aurantifolia* (Christm.) Swingle, *Planta,* 165, 142, 1985.

20. **Gmitter, F. G., Jr, and Moore, G. A.,** Plant regeneration from undeveloped ovules and embryogenic calli of *citrus:* embryo production, germination, and plant survival, *Plant Cell Tissue Organ Cult.,* 6, 139, 1986.

21. **Perez-Bermudez, P., Cornejo, M. J., and Segura, J.,** Pollen plant formation from anther cultures of *Digitalis obscura* L., *Plant Cell Tissue Organ Cult.,* 5, 63, 1985.

22. **Cappadoica, M., Chretien, L., and Laublin, G.,** Production of haploids in *Gerbera jamesonii* via ovule culture: influence of fall versus spring sampling on callus formation and shoot regeneration, *Can. J. Bot.,* 66, 1107, 1988.

23. **Yang, H. Y., Yan, H., and Zhou, C.,** *In vitro* production of haploids in *Helianthus,* in *Biotechnology in Agriculture and Forestry,* Vol. 10: *Legumes and Oilseed Crops I,* Bajaj, Y. P. S., Ed., Springer-Verlag, Berlin, 1990, 472.

24. **Ziauddin, A., Simion, E., and Kasha, K. J.,** Improved plant regeneration from shed microspore culture in barley (*Hordeum vulgare* L.) cv. Igri, *Plant Cell Rep.,* 9, 69, 1990.

25. **Piccirilli, M. and Arcioni, S.,** Haploid plants regenerated via anther culture in wild barley (*Hordeum spontaneum* C. Kock), *Plant Cell Rep.,* 10, 273, 1991.

26. **Canhoto, J. M., Ludovina, M., Guimaraes, S., and Cruz, G. S.,** *In vitro* induction of haploid, diploid and triploid plantlets by anther culture of *Iochroma warscewiczii* Regel, *Plant Cell Tissue Organ Cult.,* 21, 171, 1990.

27. **Nichterlein, K., Umbach, H., and Friedt, W.,** Genotypic and exogenous factors affecting shoot regeneration from anther callus of linseed (*Linum usitatissimum* L.), *Euphytica,* 58, 157, 1991.

28. **Olesen, A., Anderson, S. B., and Due, I. K.,** Anther culture response in perennial ryegrass (*Lolium perenne* L.), *Plant Breed.,* 101, 60, 1988.

29. **Shoukang, L., Dongfeng, J., and Jun, Q.,** *In vitro* production of haploid plants from mulberry (*Morus*) anther culture, *Sci. Sin.,* 30, 853, 1987.

30. **Chen, C., Huang, C., and To, K.,** Anther cultures of four diploid *Nicotiana* species and chromosome numbers of regenerated plants, *Bot. Bull. Acad. Sin.,* 26, 147, 1985.

31. **Lespinasse, R., DePaepe, R., and Koulou, A.,** Induction of B chromosome formation in androgenetic lines of *Nicotiana sylvestris, Caryologia,* 40, 327, 1987.

32. **Ostrem, J. A., Litton, C. C., and Collins, G. B.,** Isolation of dark tobacco breeding lines differing in alkaloid concentration via anther derived haploids, *Z. Pflanzenzuecht.,* 96, 224, 1986.

33. **Quiren, C., Zhenhua, Z., and Yuanhua, G.,** Cytogenetical analysis on aneuploids obtained from pollenclones of rice (*Oryza sativa* L.), *Theor. Appl. Genet.,* 71, 506, 1985.

34. **Dieu, P. and Dunwell, J. M.,** Anther culture with different genotypes of opium poppy (*Papaver somniferum* L.): effect of cold treatment, *Plant Cell Tissue Organ Cult.,* 12, 263, 1988.

35. **Rao, P. V. L. and De, D. N.,** Haploid plants from *in vitro* anther culture of the leguminous tree, *Peltophorum pterocarpum* (DC) K. Hayne (Copper pod), *Plant Cell Tissue Organ Cult.,* 11, 167, 1987.

36. **Racquin, C., Cornu, A., Farcy, E., Maizonnier, D., Pelletier, G., and Vedel, F.,** Nucleus substitution between *Petunia* species using gamma ray-induced androgenesis, *Theor. Appl. Genet.,* 78, 337, 1989.

37. **Stoehr, M. U. and Zsuffa, L.,** Induction of haploids in *Populus maximowiczii* via embryogenic callus, *Plant Cell Tissue Organ Cult.,* 23, 49, 1990.

38. **Hyun, S. K., Kim, J. H., Noh, E. W., and Park, J. I.,** Induction of haploid plants of *Populus* species, in *Plant Tissue Culture and Its Agricultural Applications,* Withers, L. A. and Alderson, P. G., Eds., Butterworths, London, 1986, 413.

39. **Chakravarty, B. and Sen, S.,** Regeneration through somatic embryogenesis from anther explants of *Scillia indica* (Roxb.) Baker, *Plant Cell Tissue Organ Cult.,* 19, 71, 1989.

40. **Flehinghaus, T., Deimling, S., and Geiger, H. H.,** Methodical improvements in rye anther culture, *Plant Cell Rep.,* 10, 397, 1991.

41. **Tsay, H. S., Yeh, C. C., and Hsu, J. Y.,** Embryogenesis and plant regeneration from anther culture of bamboo (*Sinocalamus latiflora* (Munro) McClure), *Plant Cell Rep.,* 9, 349, 1990.

42. **Rivard, S. R., Cappadocia, M., Vincent, G., Brisson, N., and Landry, B. S.,** Restriction fragment length polymorphism (RFLP) analyses of plants produced by *in vitro* anther culture of *Solanum chacoense* Bitt., *Theor. Appl. Genet.,* 78, 49, 1989.

43. **Powell, W. and Uhrig, H.,** Anther culture of Solanum genotypes, *Plant Cell Tissue Organ Cult.,* 11, 13, 1987.

44. **Tiainen, T.,** The role of ethylene and reducing agents on anther culture response of tetraploid potato (*Solanum tuberosum* L.), *Plant Cell Rep.,* 10, 604, 1992.

45. **Wen, F. S., Sorrenson, E. L., Barnett, F. L., and Liang, G. H.,** Callus induction and plant regeneration from anther and inflorescence culture of Sorghum, *Euphytica,* 52, 177, 1991.

46. **Hassawi, D. S. and Liang, G. H.,** Effect of cultivar, incubation temperature, and stage of microspore development on anther culture in wheat and triticale, *Plant Breed.,* 105, 332, 1990.

47. **Petolino, J. F. and Thompson, S. A.,** Genetic analysis of anther culture response in maize, *Theor. Appl. Genet.,* 74, 284, 1987.

48. **Dunwell, J. M.,** Pollen, ovule and embryo culture as tools in plant breeding, in *Plant Tissue Culture and its Agricultural Applications,* Withers, L. A. and Alderson, P. G., Eds., Butterworths, London, 1986, 357.

49. **Kasha, K. J., Ziauddin, A., and Choo, U.-H.,** Haploids in cereal improvement: anther and microspore culture, in *Gene Manipulation in Plant Improvement II,* Gustafson, J. P., Ed., Plenum Press, New York, 1990, 213.

50. **Siebel, J. and Pauls, K. P.,** A comparison of anther and microspore culture as a breeding tool in *Brassica napus, Theor. Appl. Genet.,* 78, 473, 1989.

51. **Huang, B. and Keller, W. A.,** Microspore culture technology, *J. Tissue Cult. Methods,* 12, 171, 1989.

52. **Heberle-Bors, E.,** *In vitro* haploid formation from pollen: a critical review, *Theor. Appl. Genet.,* 71, 361, 1985.

53. **Yang, H. Y. and Zhou, C.,** *In vitro* gynogenesis, in *Plant Tissue Culture Applications and Limitations,* Bhojwani, S. S., Ed., Elsevier, New York, 1990, 242.

54. **Costillo, A. M. and Cistué, L.,** Production of gynogenetic haploids of *Hordeum vulgare* L., *Plant Cell Rep.,* 12, 139, 1993.

55. **Kasha, K. J. and Reinbergs, E.,** Recent developments in the production and utilization of haploids in barley, in *Barley Genetics IV, Proc. 4th Int. Barley Genet. Symp.,* Asher, M. J. C., Ed., Edinburgh, 1981, 655.

56. **Snape, J. W., Chapman, V., Moss, V. J., Blanchard, C. E., and Miller, T. E.,** The crossabilities of wheat varieties with *Hordeum bulbosum, Heredity,* 42, 291, 1979.

57. **Sitch, L. A. and Snape, J. W.,** Doubled haploid production in winter wheat and triticale genotypes using the *Hordeum bulbosum* system, *Euphytica,* 35, 1045, 1986.

58. **Sitch, L. A. and Snape, J. W.,** The influence of the *Hordeum bulbosum* and the wheat genotypes on haploid production in wheat *(Triticum aestivum), Z. Pflanzenzuecht.,* 96, 304, 1986.

59. **Inagaki, M.,** Three steps in producing doubled haploids of wheat through the *bulbosum* technique, in *Proc. 7th Int. Wheat Genet. Symp.* Vol. 2, Miller, T. E. and Koebner, R. M. D., Eds., Int. Plant Science Research, Cambridge, England, 1988, 1109.

60. **Hagberg, G. and Hagberg, A.,** High frequency of spontaneous haploids in the progeny of an induced mutation in barley, *Hereditas,* 93, 341, 1980.

61. **Houghas, R. W., Peloquin, S. J., Gabert, A. C., and Ross, K. W.,** Haploids of the common potato, *J. Hered.,* 49, 103, 1958.

62. **Houghas, R. W., Peloquin, S. J., and Gabert, A. C.,** Effect of seed parent and pollinator on frequency of haploids in *Solanum tuberosum, Crop Sci.,* 4, 593, 1958.

63. **Pandy, K. W. and Phung, M.,** 'Hertwig Effect' in plants: induced pathenogenesis through the use of irradiated pollen, *Theor. Appl. Genet.,* 62, 295, 1982.

64. **Lacadena, J. R.,** Spontaneous and induced parthenogenesis and androgenesis, in *Haploids in Higher Plants,* Kasha, K. J., Ed., University of Guelph, Ontario, 1974, 13.

65. **Laurie, D. A. and Bennett, M. D.,** The production of haploid wheat plants from wheat × maize crosses, *Theor. Appl. Genet.,* 76, 393, 1988.

66. **Laurie, D. A. and Reymondie, S.,** High frequencies of fertilization and haploid seedling production in crosses between commercial hexaploid wheat varieties and maize, *Plant Breed.,* 106, 182, 1991.

67. **Suenaga, K.,** An effective method of production of dihaploid wheat *(Triticum aestivum)* plants by wheat × maize *(Zea mays)* crosses, in *Int. Coll. for Overcoming Breeding Barriers,* Adachi, A., Ed., Miyazaki, Japan, 1991, 195.

68. **Laurie, D. A.,** Factors affecting fertilization frequency in *Triticum aestivum* cv Highbury × *Zea mays* cv Seneca 60 crosses, *Plant Breed.,* 103, 133, 1989a.

69. **Sitch, L. A. and Snape, J. W.,** Factors affecting haploid production in wheat using the *Hordeum bulbosum* system. I. Genotypic and environmental effects on pollen grain germination, pollen tube growth and the frequency of fertilization, *Euphytica,* 36, 483, 1987.

70. **Snape, J. W., DeBuyser, J., Henry, Y., and Simpson, E.,** A comparison of methods of haploid production in a cross of wheat *Triticum aestivum, Z. Pflanzenzuecht.,* 96, 320, 1986.

71. **Petolino, J. F., Jones, A. M., and Thompson, S. A.,** Selection for increased anther culture response in maize, *Theor. Appl. Genet.,* 76, 157, 1988.

72. **Arnison, P. G. and Keller, W. A.,** A survey of the anther culture response of *Brassica oleracea* L. cultivars grown under field conditions, *Plant Breed.,* 104, 125, 1990.

73. **Sonnino, A., Tanaka, S., Iwanaga, M., and Schilde-Rentschler, L.,** Genetic control of embryo formation in anther culture of diploid potatoes, *Plant Cell Rep.,* 8, 105, 1989.

74. **Agache, S., Bacheller, B., DeBuyser, J., Henry, Y., and Snape, J.,** Genetic control of anther culture response in wheat using aneuploid, chromosome substitution and translocation lines, *Theor. Appl. Genet.,* 77, 7, 1989.

75. **Foroughi-Wehr, B. and Zeller, F. J.,** *In vitro* microspore reaction of different German wheat cultivars, *Theor. Appl. Genet.,* 79, 77, 1990.

76. **Jones, A. M. and Petolino, J. F.,** Effects of donor plant genotype and growth environment on anther culture of soft-red winter wheat (*Triticum aestivum* L.), *Plant Cell Tissue Organ Cult.,* 8, 215, 1987.

77. **Lee, S. Y., Lee, Y. T., and Lee, M. S.,** Studies on the anther culture of *Oryza sativa* L. III. Growing environment of donor plant in anther culture — effects of photoperiod and light intensity, *The Research Reports of the Rural Development Administration,* 30, 7, 1988.

78. **Luckett, D. J. and Smithard, R. A.,** Doubled haploid production by anther culture for Australian barley breeding, *Aust. J. Agric. Res.,* 43, 67, 1992.

79. **Bjornstad, A., Opsahl-Ferstad, H.-G., and Aasmo, M.,** Effects of donor plant environment and light during incubation on anther cultures of some spring wheat *(Triticum aestivum L.) cultivars, Plant Cell Tissue Organ Cult.,* 17, 27, 1989.

80. **Kyo, M. and Harada, H.,** Studies on conditions for cell division and embryogenesis in isolated pollen culture in *Nicotiana rustica, Plant Physiol.,* 79, 90, 1985.

81. **Muller, D. and Keller, J.,** Anther culture of brussel sprouts *(Brassica oleracea)* var *gemmifera), Arch. Zuechtungsforsch.,* 20, 219, 1990.

82. **Takahata, Y. and Keller, W. A.,** High frequency embryogenesis and plant regeneration in isolated microspore culture of *Brassica oleracea* L., *Plant Sci.,* 74, 235, 1991.

83. **Kott, L. S., Polonsi, L., Ellis, B., and Beversdorf, W. D.,** Autotoxicity in isolated microspore cultures of *Brassica napus, Can. J. Bot.,* 66, 1665, 1988b.

84. **Fan, Z., Holbrook, L., and Keller, W. A.,** Isolation and enrichment of embryogenic microspores in *Brassica napus* L. by fractionation using percoll density gradients, *Proc. 7th Int. Rapeseed Congr.,* 1988, 92.

85. **Rashid, A. and Reinert, J.,** Selection of embryogenic pollen from cold-treated buds of *Nicotiana tabacum* var. Badischer Burley and their development into embryos in culture, *Protoplasma,* 105, 161, 1980.

86. **Kyo, M. and Harada, H.,** Control of the developmental pathway of tobacco pollen *in vitro, Planta,* 168, 427, 1986.

87. **Heberle-Bors, E.,** Isolated pollen culture in tobacco: plant reproductive development in a nutshell, *Sex. Plant Reprod.,* 2, 1, 1989.

88. **Johansson, L.,** Improved methods for induction of embryogenesis in anther cultures of *Solanum tuberosum, Potato Res.,* 29, 179, 1986.

89. **Lazer, M. D., Schaeffer, G. W., and Baenziger, P. S.,** The physical environment in relation to high frequency callus and plantlet development in anther culture of wheat (*Triticum aestivum* L.) cv. Chris, *J. Plant Physiol.,* 121, 103, 1985.

90. **Lazar, M. D., Schaeffer, G. W., and Baenziger, P. S.,** The effects of interactions of culture environment with genotype on wheat *(Triticum aestivum)* anther culture response, *Plant Cell Rep.,* 8, 525, 1990.

91. **Genovesi, A. D. and Collins, G. B.,** *In vitro* production of haploid plants of corn via anther culture, *Crop Sci.,* 22, 1137, 1982.

92. **Huang, B. and Sunderland, N.,** Temperature-stress pretreatment in barley anther culture, *Ann. Bot.,* 49, 77, 1982.

93. **Fitch, M. M. and Moore, P. H.,** Haploid production from anther culture of *Saccharum spontaneum* L., *Z. Pflanzenphysiol.,* 109, 197, 1983.

94. **Zaki, M. A. M. and Dickenson, H. G.,** Microspore-derived embryos in *Brassica:* the significance of division symmetry in pollen mitosis I to embryogenic development, *Sex. Plant Reprod.,* 4, 48, 1991.

95. **Klimaszewska, K. and Keller, W. A.,** The production of haploids from *Brassica hirta* Moench (*Sinapis alba* L.) anther cultures, *Z. Pflanzenphysiol.,* 109, 235, 1983.

96. **Ouyang, J.,** Induction of pollen plants in *Triticum aestivum,* in *Haploids of Higher Plants in vitro,* Hu, H. and Yang, H., Eds., Springer-Verlag, Berlin, 1986, 22.

97. **Pechan, P. M. and Keller, W. A.,** Induction of microspore embryogenesis in *Brassica napus* L. by gamma irradiation and ethanol stress, *In Vitro Cell. Dev. Biol.,* 25, 1073, 1989.

98. **Chu, C. C., Hill, R. D., and Brule-Babel, A. L.,** High frequency of pollen embryoid formation and plant regeneration in *Triticum aestivum* L. on monosaccharide containing media, *Plant Sci.,* 66, 255, 1990.

99. **Orshinsky, B. R., McGregor, L. J., Johnson, G. I. E., Hucl, P., and Kartha, K. K.,** Improved embryoid induction and green shoot regeneration from wheat anthers cultured in medium with maltose, *Plant Cell Rep.,* 9, 365, 1990.

100. **Henry, Y. and DeBuyser, J.,** Effect of the 1B/1R translocation on anther culture ability in wheat (*Triticum aestivum* L.), *Plant Cell Rep.,* 4, 307, 1985.

101. **Olsen, F. L.,** Induction of microspore embryogenesis in cultured anthers of *Hordeum vulgare*. The effects of ammonium nitrate, glutamine, and asparagine as nitrogen sources, *Carlsberg Res. Commun.*, 52, 393, 1987.

102. **Lichter, R.,** Anther culture of *Brassica napus* in a liquid culture medium, *Z. Pflanzenphysiol.*, 103, 229, 1981.

103. **Lichter, R.,** Induction of haploid plants from isolated pollen of *Brassica napus*, *Z. Pflanzenphysiol.*, 105, 427, 1982.

104. **Charne, D. G. and Beversdorf, W. D.,** Improving microspore culture as a rapeseed breeding tool: the use of auxins and cytokinins in an induction medium, *Can. J. Bot.*, 66, 1671, 1988.

105. **Biddington, N. L. and Robinson, H. T.,** Ethylene production during anther culture of brussels sprouts (*Brassica oleracea* var *gemmifera*) and its relationship with factors that affect embryo production, *Plant Cell Tissue Organ Cult.*, 25, 169, 1991.

106. **Reynolds, T. L.,** A possible role for ethylene during IAA-induced pollen embryogenesis in anther cultures of *Solanum carolinense* L., *Am. J. Bot.*, 74, 967, 1987.

107. **Cho, U.-H., and Kasha, K. J.,** Ethylene production and embryogenesis from anther cultures of barley *(Hordeum vulgare)*, *Plant Cell Rep.*, 8, 415, 1990.

108. **Pace, G. M., Ried, J. N., Ho, L. C., and Fahey, J. W.,** Anther culture of maize and the visualization of embryogenic microspores by fluorescent microscopy, *Theor. Appl. Genet.*, 73, 863, 1987.

109. **Kuhlmann, U. and Foroughi-Wehr, B.,** Production of doubled haploid lines in frequencies sufficient for barley breeding programs, *Plant Cell Rep.*, 8, 78, 1989.

110. **Calleberg, E. K., Kristjansdottir, I. S., and Johansson, L. B.,** Anther cultures of tetraploid *Solanum* genotypes — the influence of gelling agents and correlations between incubation temperature and pollen germination temperature, *Plant Cell Tissue Organ Cult.*, 19, 189, 1989.

111. **Ziegler, G., Dressler, K., and Hess, D.,** Investigations on the anther culturability of four German spring wheat cultivars and the influence of light on regeneration of green vs. albino plants, *Plant Breed.*, 105, 40, 1990.

112. **Hunter, C. P.,** The effect of anther orientation on the production of microspore-derived embryoids and plants of *Hordeum vulgare* cv. Sabarlis, *Plant Cell Rep.*, 4, 267, 1985.

113. **Ulrich, A., Furtan, W. H., and Downey, R. K.,** Biotechnology and rapeseed breeding: some economic considerations, *Sci. Counc. Can. Rep.*, 67pp, 1984.

114. **Mathias, R. and Robbelen, G.,** Effective diploidization of microspore-derived haploids of rape (*Brassica napus* L.) by *in vitro* colchicine treatment, *Plant Breed.,* 106, 82, 1991.

115. **Chen, J. L. and Beversdorf, W. D.,** Production of spontaneous diploid lines from isolated microspores following cryopreservation of spring rapeseed (*Brassica napus* L.), *Plant Breed.,* 108, 324, 1992.

116. **Powell, W., Caligari, D. D. S., and Dunwell, J. M.,** Field performance of lines derived from haploid and diploid tissues of *Hordeum vulgare, Theor. Appl. Genet.,* 72, 458, 1986.

117. **Burk, L. G. and Matzinger, D. F.,** Variation among anther-derived doubled haploids from an inbred line of tobacco, *J. Hered.,* 67, 381, 1976.

118. **Sangwan, R. S. and Sangwan-Norreel, B. S.,** Anther and pollen culture, in *Plant Tissue Culture: Applications and Limitations,* Bhojwani, S.S., Ed., Elsevier, Oxford, 1990, 220.

119. **Friedt, W. and Foroughi-Wehr, B.,** Field performance of androgenetic doubled haploid spring barley and F_1 hybrids, *Z. Pflanzenzuecht.,* 90, 117, 1983.

120. **Charmet, G. and Branlard, G.,** A comparison of androgenetic doubled haploid and single seed descent lines of Triticale, *Theor. Appl. Genet.,* 71, 193, 1985.

121. **Lichter, R., DeGroat, E., Fiebeg, D., Schweiger, R., and Gland, A.,** Glucosinolates determined by HPLC in the seeds of microspore derived homozygous lines of rapeseed (*Brassica napus* L.), *Plant Breed.,* 100, 209, 1988.

122. **Hoffman, F., Thomas, E., and Wenzel, G.,** Anther culture as a tool in Rape. II. Progeny analyses of androgenetic lines and induced mutants from haploid cultures, *Theor. Appl. Genet.,* 61, 225, 1982.

123. **Ho, K. M. and Jones, G. E.,** 'Mingo' barley, *Can. J. Plant Sci.,* 60, 279, 1980.

124. **Campbell, K. W., Brown, R. I., and Ho, K. M.,** 'Rodeo' barley, *Can. J. Plant Sci.,* 64, 203, 1984.

125. **Baenziger, P. S., Kudirka, D. T., Schaeffer, G. W., and Lazar, M. D.,** The significance of doubled haploid variation, in *Gene Manipulation and Plant Improvement,* Gustafson, J. P., Plenum Press, New York, 1984, 385.

126. **Nakamura, A. T., Yamada, M., Oka, Y., Tatemichi, K., Egushi, T., Ayabe, T., and Kobayashi, K.,** Studies on the haploid method of breeding by anther culture in tobacco. V. Breeding of mild flue-cured variety F211 by haploid method, *Bull. Iwata Tobacco Exp. Stn.,* 7, 29, 1975.

127. **Anon.,** Success of breeding the new tobacco cultivar 'Tan-Yuh no. 1', *Acta Bot. Sin.,* 16, 300, 1974.

128. **Chaplin, J. F., Burk, L. G., Gooding, G. V., and Powell, N. T.,** Registration of NC 744 tobacco germplasm (Reg No GP 18), *Crop Sci.,* 20, 677, 1980.

129. **Chaplin, J. F. and Burk, L. G.,** Registration of LMAFC 34 tobacco germplasm, *Crop Sci.,* 24, 1220, 1984.

130. **Anon.,** New rice varieties Hua Yu 1 and Hua Yu 2 developed from anther culture, *Acta Genet. Sin.,* 3, 19, 1976.

131. **Daofen, H., Zhendong, Y., Yunlian, T., and Jianping, L.,** Jinghua No. 1 — a winter wheat variety derived from pollen sporophyte, *Sci. Sin.,* 29, 733, 1986.

132. **DeBuyser, J., Henry, Y., Lonnet, P., Hertzog, R., and Hespel, A.,** 'Florin': a doubled haploid wheat variety developed by the anther culture method, *Plant Breed.,* 98, 53, 1987.

133. **Huang, C. S., Tsay, H. S., Chern, C. G., Chen, C. C., Yeh, C. C., and Tseng, T. H.,** Japonica rice breeding using anther culture, *J. Agric. Res. China,* 37, 1, 1988.

134. **Hu, H. and Huang, B.,** Applications of pollen-derived plants to crop improvement, *Int. Rev. Cytol.,* 107, 293, 1987.

135. **Choo, T. M., Reinbergs, E., and Kasha, K. J.,** Use of haploids in breeding barley, *Plant Breed. Rev.,* 3, 219, 1985.

136. **Rajhathy, T.,** Haploid flax revisited, *Z. Pflanzenzuecht.,* 76, 1, 1976.

137. **Henderson, C. A. P. and Pauls, K. P.,** The use of haploidy to develop plants that express several recessive traits using lightseeded canola *(Brassica napus)* as an example, *Theor. Appl. Genet.,* 83, 476, 1992.

138. **Siebel, J. and Pauls, K. P.,** Inheritance pattern of erucic acid content in populations of *Brassica napus* microspore-derived spontaneous diploids, *Theor. Appl. Genet.,* 77, 489, 1989.

139. **Choo, T. M. and Reinbergs, E.,** Doubled haploids for estimating genetic variances in the presence of linkage and gene association, *Theor. Appl. Genet.,* 55, 129, 1979.

140. **Wang, X. Z. and Hu, H.,** Chromosome constitution of plants derived from pollen of hexaploid triticale × common wheat F_1 hybrid, *Theor. Appl. Genet.,* 70, 92, 1985.

141. **Miao, Z. H., Zhuang, J. J., and Hu, H.,** Expression of various gametic types in pollen plants regenerated from hybrids between *Triticum-Agropyron* and wheat, *Theor. Appl. Genet.,* 75, 485, 1988.

142. **Tao, Y. Z. and Hu, H.,** Recombination of R-D chromosome in pollen plants cultured from hybrid of 6x triticale × common wheat, *Theor. Appl. Genet.*, 77, 899, 1989.

143. **Straub, J.,** Theoretical aspects of haploid techniques, in *Plant Tissue Culture and its Biotechnological Applications*, Barz, W., Reinhard, E., Zenk, M. H., Eds., Springer-Verlag, Berlin, 1977, 334.

144. **Jansky, S. H., Yerk, G. L., and Peloquin, S. J.,** The use of potato haploids to put 2x wild species germplasm into a usable form, *Plant Breed.*, 104, 290, 1990.

145. **Chen, J. L. and Beversdorf, W. D.,** A comparison of traditional and haploid-derived breeding populations of oilseed rape (*Brassica napus* L.) for fatty acid composition of the seed oil, *Euphytica*, 51, 59, 1990.

146. **Scarth, R., Seguin-Swartz, G., and Rakow, G. F. W.,** Application of doubled haploidy to *Brassica* breeding, in *Proc. 8th Int. Rapeseed Congr.*, McGregor, D. I., Ed., 1991, 1449.

147. **Brown, J. S. and Wernsman, E. A.,** Nature of reduced productivity of anther-derived dihaploid lines of flue-cured tobacco, *Crop Sci.*, 22, 1, 1982.

148. **Zivy, M., Devaux, P., Blaisonneau, J., Jean, R., and Thiellement, H.,** Segregation distortion and linkage studies in microspore-derived doubled haploid lines of *Hordeum vulgare* L., *Theor. Appl. Genet.*, 83, 919, 1992.

149. **Guiderdoni, E.,** Gametic selection in anther culture of rice (*Oryza sativa* L.), *Theor. Appl. Genet.*, 81, 406, 1991.

150. **Sacristan, M. D.,** Resistance responses to *Phoma lingam* of plants regenerated from selected cell and embryogenic cultures of haploid *Brassica napus*, *Appl. Genet.*, 61, 193, 1982.

151. **Carlson, P. S.,** Methionine sulfoximine-resistant mutants of tobacco, *Science*, 180, 1366, 1973.

152. **Creason, G. L. and Chaleff, R. S.,** A second mutation enhances resistance of a tobacco mutant to sulfonylurea herbicides, *Theor. Appl. Genet.*, 76, 177, 1988.

153. **Chaleff, R. S. and Ray, T. B.,** Herbicide resistant mutants from tobacco cell cultures, *Science*, 223, 1148, 1984.

154. **Negrutiu, I., De Brouwer, D., Dirks, R., and Jacobs, M.,** Amino acid auxotrophs from protoplast cultures of *Nicotiana plumbaginifolia* Viviani, I. BUdR enrichment selection, plant regeneration, and general characterization, *Mol. Gen. Genet.*, 14, 330, 1985.

155. **Maliga, P.,** Isolation and characterization of mutants in plant cell cultures, *Annu. Rev. Plant Physiol.,* 35, 519, 1984.

156. **Schaeffer, G. W. and Sharpe, F. T., Jr.,** Lysine in seed protein of S-aminoethyl-cysteine resistant anther-derived tissue culture of rice, *In Vitro,* 17, 345, 1981.

157. **Schaeffer, G. W. and Sharpe, F. T., Jr.,** Mutations and cell selection: genetic variation for improved protein in rice, in *Proc. Beltsville Symp. Agric. Res.,* Owens, L. D., Ed., Rowman and Allanheld, NJ, 1983, 257.

158. **Jacobsen, E., Visser, R. G. F., and Wijbrandi, J.,** Phenylalanine and tyrosine accumulating cell lines of a dihaploid potato selected by resistance to 5-methyltryptophan, *Plant Cell Rep.,* 4, 151, 1985.

159. **Sidorov, V. A. and Maliga, P.,** Fusion complementation analysis of auxotrophic and chlorophyll-deficient lines isolated in haploid *Nicotiana plumbaginifolia* protoplast cultures, *Mol. Gen. Genet.,* 186, 328, 1982.

160. **Shimamoto, K. and King, P. J.,** Isolation of a histidine auxotroph of *Hyoscyamus muticus* during attempts to apply BUdR enrichment, *Mol. Gen. Genet.,* 189, 69, 1983.

161. **Gebhardt, C., Schnebli, V., and King, P. J.,** Isolation of biochemical mutants using haploid mesophyll protoplasts of *Hyoscyamus muticus.* II. Auxotrophic and temperature-sensitive clones, *Planta,* 153, 81, 1981.

162. **Steffen, A. and Schieder, O.,** Biochemical and genetical characterization of nitrate reductase deficient mutants of *Petunia, Plant Cell Rep.,* 3, 139, 1984.

163. **Blondstein, A. D., Vahala, T., Fracheboud, Y., and King, P. J.,** Temperature-sensitive plant mutants isolated *in vitro, Mol. Gen. Genet.,* 21, 252, 1988.

164. **Swanson, E. B., Coumans, M. P., and Brown, G. L.,** The characterization of herbicide tolerant plants in *Brassica napus* L. after *in vitro* selection of microspore and protoplasts, *Plant Cell Rep.,* 7, 83, 1988.

165. **Swanson, E. B., Herrgesell, M. J., and Arnaldo, M.,** Microspore mutagenesis and selection: canola plants with field tolerance to the imidazolinones, *Theor. Appl. Genet.,* 78, 520, 1989.

166. **Turner, J. and Facciotti, D.,** High oleic acid *Brassica napus* from mutagenized microspores, in *Proc. 6th Crucifer Genetics Workshop,* McFerson, J. R., Kresovich, S., and Dwyer, S. G., Eds., USDA ARS, Geneva, NY, 1990, 24.

167. **Huang, B., Swanson, E. B., Baszczynski, C. L., Macrae, W. D., Barbour, E., Armavil, V., Woke, L., Arnaldo, M., Rozakis, S., Westecott, M., Keats, R. F., and Kemble, R.,** Application of microspore culture to canola improvement, in *Proc. 8th Int. Rapeseed Congress,* McGregor, D. I., Ed., 298, 1991.

168. **Huang, B.,** Genetic manipulation of microspores and microspore-derived embryos, *In Vitro, Cell Dev. Biol.,* 28, 53, 1992.

169. **Nitsch, J. P., Nitsch, C., and Pereau-Leroy, P.,** Obtention de mutant a partier de nicotiana haploides issus de grains de pollen, *C.R. Acad. Sci. Paris Ser.,* 269, 1650, 1969.

170. **Polsoni, L., Kott, L. S., and Beversdorf, W. D.,** Large-scale microspore culture technique for mutation-selection studies in *Brassica napus, Can. J. Bot.,* 66, 1681, 1988.

171. **MacDonald, M. V., Hadwiger, M. A., Aslam, F. N., and Ingram, D. S.,** The enhancement of anther culture efficiency in *Brassica napus* ssp *oleifera* Metzg (Sinsk) using low doses of gamma irradiation, *New Phytol.,* 110, 101, 1988.

172. **Medrano, H., Primo-Millo, E., and Guerri, J.,** Ethyl-methane-sulfonate effects on anther cultures of *Nicotiana tabacum, Euphytica,* 35, 161, 1986.

173. **Zhang, Y. X., Bouvier, L., and Lespinasse, Y.,** Microspore embryogenesis induced by low gamma dose irradiation in apple, *Plant Breed.,* 108, 173, 1992.

174. **Przewozny, T., Schieder, O., and Wenzel, G.,** Induced mutants from dihaploid potatoes after pollen mother cell treatment, *Theor. Appl. Genet.,* 58, 145, 1980.

175. **Larkin, P. J. and Scowcroft, W. R.,** Somaclonal variation — a novel source of variability from cell cultures for plant improvement, *Theor. Appl. Genet.,* 60, 197, 1981.

176. **Morrison, R. A. and Evans, D. A.,** Gametoclonal variation, *Plant Breed. Rev.,* 5, 359, 1987.

177. **Simantel, G. M. and Ross, J. G.,** Colchicine-induced somatic chromosome reduction in Sorghum, *J. Hered.,* 553, 1964.

178. **Hague, L. M. and Jones, R. N.,** Cytogenetics of *Lolium perenne:* colchicine-induced variation in diploids, *Theor. Appl. Genet.,* 74, 233, 1987.

179. **Wang, Y. and Hu, H.,** Gamete composition and chromosome variation in pollen-derived plants from octoploid triticale × common wheat hybrids, *Theor. Appl. Genet.,* 85, 681, 1993.

180. **Reed, S. M. and Wernsman, E. A.,** DNA amplification among anther-derived doubled haploid lines of tobacco and its relationship to agronomic performance, *Crop Sci.,* 29, 1072, 1989.

181. **Dhillon, S. S., Wernsman, E. A., and Miksche, J. P.,** Evaluation of nuclear DNA content and heterochromatin changes in anther-derived dihaploids of tobacco *(Nicotiana tabacum),* CV Coker 139, *Can. J. Genet. Cytol.,* 25, 169, 1983.

182. **Chu, Q., Zhang, Z., and Gao, Y.,** Cytological analysis on aneuploids obtained from pollen clones of rice *(Oryza sativa), Theor. Appl. Genet.,* 71, 506, 1985.

183. **Kumashiro, T. and Oinuma, T.,** Comparison of genetic variability among anther-derived and ovule-derived doubled haploid lines of tobacco, *Jpn. J. Breed.,* 35, 301, 1985.

184. **San Noeum, L. H. and Ahmadi, N.,** Variability of doubled haploids from *in vitro* androgenesis and gynogenesis, in *Variability in Plants Regenerated from Tissue Culture,* Earle, E. and Demarly, Y., Eds., Praeger, New York, 1983, 273.

185. **Witherspoon, W. D., Jr., Wernsman, E. A., Gooding, G. V., Jr., and Rufty, R. C.,** Characterization of a gametoclonal variant controlling virus resistance in tobacco, *Theor. Appl. Genet.,* 81, 1, 1991.

186. **Nichols, W. A. and Rufty, R. C.,** Anther culture as a probable source of resistance to tobacco black shank caused by *Phytophthora parasitica* var *nicotianae, Theor. Appl. Genet.,* 84, 473, 1992.

187. **Sacristan, M. D.,** Selection for disease resistance in *Brassica* cultures, *Hered. Suppl.,* 3, 57, 1985.

188. **Newsholme, D. M., MacDonald, M. V., and Ingram, D. S.,** Studies of selection *in vitro* for novel resistance to phytotoxic products by *Leptosphaeria maculans* (Desm.) Ces and de Not, in secondary embryogenic lines of *Brassica napus* ssp. *oleifera* (Metzg.), Sinsk, Winter oilseed rape, *New Phytol.,* 113, 117, 1989.

189. **Hodgkin, T. and MacDonald, M. V.,** The effect of a phytotoxin from *Alternaria brassicicola* on *Brassica* pollen, *New Phytol.,* 104, 631, 1986.

190. **Friedt, W. and Foroughi-Wehr, B.,** Rapid production of recombinant barley yellow mosaic virus resistant *Hordeum vulgare* lines by anther culture, *Theor. Appl. Genet.,* 67, 377, 1984.

191. **Voorrips, R. E. and Visser, D. L.,** Doubled haploid lines with clubroot resistance in *Brassica oleracea,* in *Proc. 6th Crucifer Genet. Workshop,* McFerson, J. R., Kresovich, S., and Dwyer, S., Eds., USDA ARS, Geneva, NY, 1990, 40.

192. **Burk, L. G. and Chaplain, J. F.,** Variation among anther-derived haploids from multiple disease resistant tobacco hybrids, *Crop Sci.,* 20, 334, 1980.

193. **Schaeffer, G. W., Sharpe, F. T., and Cregan, P. B.,** Variation for improved protein and yield from rice anther culture, *Theor. Appl. Genet.,* 67, 313, 1984.

194. **Deaton, W. R., Legg, P. D., and Collins, G. B.,** A comparison of burley tobacco doubled haploid lines with the source inbred cultivars, *Theor. Appl. Genet.,* 62, 69, 1982.

195. **Collins, G. B., Legg, P. B., and Kasperbauer, M. J.,** Use of anther-derived haploids in *Nicotiana*. I. Isolation of breeding lines differing in total alkaloid content, *Crop Sci.,* 14, 77, 1974.

196. **Lashermes, P.,** Screening for stress tolerant genotypes via microspore *in vitro* culture, in *Physiological Breed Winter Cereal for Stressed Mediterranean Environments,* INRA, Montpelier, France, 1989, 463.

197. **Neuhaus, G., Spangenberg, G., Mittelsten-Sheid, O., and Schweiger, G.,** Transgenic rapeseed plants obtained by the microinjection of DNA into microspore-derived embryoids, *Theor. Appl. Genet.,* 75, 30, 1987.

198. **Oelck, M. M., Phan, C. V., Eckes, P., Donn, G., Rakow, G., and Keller, W. A.,** Field resistance of canola transformants (*Brassica napus* L.) to ignite (Phosphinothricin), in *Proc. 8th Int. Rapeseed Congr.,* McGregor, D. I., Ed., 1991, 293.

199. **Swanson, E. B. and Erickson, L. R.,** Haploid transformation in *Brassica napus* using an octopine-producing strain of *Agrobacterium tumefaciens, Theor. Appl. Genet.,* 78, 831, 1989.

200. **Kuhlmann, U., Foroughi-Wehr, B., Garner, A., and Wenzel, G.,** Improved culture system for microspore of barley to become a target for DNA uptake, *Plant Breed.,* 107, 165, 1991.

201. **Fennell, A. and Hauptmann, R.,** Electroporation and PEG delivery of DNA into maize microspores, *Plant Cell Rep.,* 11, 567, 1992.

202. **D'Halluin, K., Bonne, E., Bossut, M., DeBuckeleer, M., and Leemans, J.,** Transgenic maize plants by tissue electroporation, *Plant Cell,* 4, 1495, 1992.

203. **Topper, R., Gronenborn, B., Shell, J., and Steinbiss, H. H.,** Uptake and transient expression of chimeric genes in seed-derived embryos, *Plant Cell,* 1, 133, 1989.

204. **Pechan, P. M.,** Successful cocultivation of *Brassica napus* microspores and proembryos with *Agrobacterium, Plant Cell Rep.,* 8, 387, 1989.

205. **Guiderdoni, E. and Chair, H.,** Plant regeneration from haploid cell suspension-derived protoplasts of Mediterranean rice (*Oryza sativa* L. cv. Miara), *Plant Cell Rep.,* 11, 618, 1992.

206. **Su, R.-C., Rudert, M. L., and Hodges, T. K.,** Fertile indica and japonica rice plants regenerated from protoplasts isolated from embryogenic haploid suspension cultures, *Plant Cell Rep.,* 12, 45, 1992.

207. **Chuong, P. V., Beversdorf, W. D., Powell, A. D., and Pauls, K. P.,** The use of haploid protoplast fusion to combine cytoplasmic atrazine resistance and cytoplasmic male sterility in *Brassica napus, Plant Cell Tissue Organ Cult.,* 12, 181, 1988.

208. **Chuong, P. V., Beversdorf, W. D., Powell, A. D., and Pauls, K. P.,** Somatic transfer of cytoplasmic traits in *Brassica napus* L. by haploid protoplast fusion, *Mol. Gen. Genet.,* 211, 197, 1988.

209. **Bajaj, Y. P. S.,** Haploid protoplasts, in *International Review of Cytology,* Supplement 16, Giles, K. L., Ed., Academic Press, New York, 1983, 113.

210. **Pirrie, A. and Power, J. B.,** The production of fertile triploid somatic hybrid plants (*Nicotiana glutinosa* (N) + *N. tabacum* (2N) via gametic:somatic protoplast fusion, *Theor. Appl. Genet.,* 72, 48, 1986.

211. **Pental, D., Mukhopadhyay, A., Grover, A., and Pradham, A. K.,** A selection method for the synthesis of triploid hybrids by fusion of microspore protoplasts (n) with somatic cell protoplasts (2n), *Theor. Appl. Genet.,* 76, 237, 1988.

212. **Pental, D. and Cocking, E. C.,** Some theoretical and practical possibilities of plant genetic manipulation using protoplasts, *Hereditas (Suppl.),* 3, 83, 1985.

213. **Tanaka, I., Kitazume, C., and Ito, M.,** The isolation and culture of lily pollen protoplasts, *Plant Sci.,* 50, 205, 1987.

214. **Weaver, M. L., Breda, V., Gaffield, W., and Timm,. H.,** Nonenzymic release of intact protoplasts from mature pollen of bean, *J. Am. Soc. Hortic. Sci.,* 115, 640, 1990.

215. **Fellner, M. and Havrânek, P.,** Isolation of *Allium* pollen protoplasts, *Plant Cell Tissue Organ Cult.,* 29, 275, 1992.

216. **Pechan, P. M., Keller, W. A., Mandy, F., and Bergeron, M.,** Selection of *Brassica napus* L. embryogenic microspores by flow sorting, *Plant Cell Rep.,* 7, 396, 1988.

217. **Pechan, P. M. and Keller, W. A.,** Identification of potentially embryogenic microspores in *Brassica napus, Physiol. Plant.,* 74, 377, 1988.

218. **Crouch, M. L.,** Nonzygotic embryos of *Brassica napus* L. contain embryo-specific storage proteins, *Planta,* 156, 520, 1982.

219. **Pomeroy, M. K., Kramer, J. K. G., Hunt, D. J., and Keller, W. A.,** Fatty acid changes during development of zygotic and microspore-derived embryos of *Brassica napus, Physiol. Plant.,* 81, 447, 1991.

220. **Pomeroy, M. K., Sparace, S. A., and Keller, W. A.,** Fatty acid and triacylglycerol biosynthesis in microspore embryos of *Brassica napus,* in *Proc. 8th Int. Rapeseed Congr.,* McGregor, D. I., Ed., 1991, 154.

221. **Taylor, D. C., Weber, N., Underhill, E. W., Pomeroy, M. K., Keller, W. A., Scowcroft, W. R., Wilen, R. W., Moloney, M. M., and Holbrook, L. A.,** Storage protein regulation and lipid accumulation of microspore embryos of *Brassica napus* L., *Planta,* 181, 18, 1990.

222. **Wiberg, E., Rahlen, L., Hellman, M., Tillberg, E., Glimelius, K., and Stymne, S.,** The microspore-derived embryo of *Brassica napus* L. as a tool for studying embryo-specific lipid biogenesis and regulation of oil quality, *Theor. Appl. Genet.,* 82, 515, 1991.

223. **Holbrook, L. A., Van Rooijen, G. J. H., Wilen, R. W., and Moloney, M. M.,** Oil-body proteins in microspore-derived embryos of *Brassica napus*. Hormonal, osmotic and developmental regulation of synthesis, *Plant Physiol.,* 97, 1051, 1991.

224. **Gruber, S. and Röbbelen, G.,** Fatty acid synthesis in microspore-derived embryoids of rapeseed *(Brassica napus),* in *Proc. 8th Int. Rapeseed Congr.,* McGregor, D. I., Ed., 1991, 1818.

225. **Weselake, R. J., Taylor, D. C., Pomeroy, M. K., Lewson, S. L., and Underhill, E. W.,** Properties of diacylglycerol acyltransferase from microspore-derived embryos of *Brassica napus* L., *Phytochemistry,* 30, 3533, 1991.

226. **Eikenberry, E. J., Choung, P. V., Esser, J., Romero, J., and Ram, R.,** Maturation, desiccation, germination and storage lipid accumulation in microspore embryos of *Brassica napus* L., in *Proc. 8th Int. Rapeseed Congr.,* McGregor, D. I., Ed., 1991, 1809.

227. **Taylor, D. C., Weber, N., Hoffe, L. R., Underhill, E. W., and Pomeroy, M. K.,** Formation of trierucoylglycerol *(Trierucin)* from 1,2 dierucoylglycerol by a homogenate of microspore-derived embryos of *Brassica napus* L., *J. Am. Oil Chem. Soc.,* 69, 355, 1992.

228. **Taylor, D. C., Weber, N., Barton, D. L., Underhill, E. W., Hoffe, L. R., Weselake, R. J., and Pomeroy, M. K.,** Triacylglycerol bioassembly in microspore-derived embryos of *Brassica napus* L. cv. Reston, *Plant Physiol.,* 97, 65, 1991.

229. **Weber, N., Taylor, D. C., and Underhill, E. W.,** Biosynthesis of storage lipids in plant cell and embryo cultures, *Adv. Biochem. Eng. Biotechnol.,* 45, 99, 1992.

230. **Weber, N. and Taylor, D. C.,** Biosynthesis of triacylglycerols in plant cell and embryo cultures. Their significance in the breeding of oilseeds, in *Progress in Plant Cellular and Molecular Biology,* Nijkamp, H. J. J., Van Der Plas, L. H. W., and Van Aartrijk, J., Eds., Kluwer Academic, Dordrecht, 1990, 324.

231. **Taylor, D. C., Ferrie, A. M. R., Keller, W. A., Giblin, E. M., Pass, E. U., and Mackenzie, S. L.,** Bioassembly of acyllipids in microspore-derived embryos of *Brassica campestris, Plant Cell Rep.*, 12, 375, 1993.

232. **Holbrook, L. A., Scowcroft, W. R., Taylor, D. C., Pomeroy, M. R., Wilen, R. W., and Moloney, M. M.,** Microspore-derived embryos: a tool for studies in regulation of gene expression in zygotic embryos, in *Progress in Plant Cellular and Molecular Biology,* Nijkamp, N. J. J., Van Der Plas, L. H. W., and Van Aartrijk, J., Eds., Kluwer Academic, Dordrecht, 1990, 402.

233. **Taylor, D., Weber, N., Barton, D., Underhill, E., and Pomeroy, K.,** Biosynthesis of triacylglycerols containing eruicic acid in microspore-derived embryos of *Brassica napus,* in *Plant Lipid Biochemistry, Structure and Utilization,* Quinn, P. J. and Harwood, J. L., Eds., Portland Press, London, 1990, 210.

234. **Holbrook, L. A., Magus, J. R., and Taylor, D. C.,** ABA induction of elongase activity, biosynthesis and accumulation of very long chain monounsaturated fatty acid and oil body proteins in microspore-derived embryos of *Brassica napus* L. cv. Reston, *Plant Sci.,* 84, 99, 1992.

235. **Taylor, D. C., Barton, D. L., Roux, K. P., Mackenzie, S. I., Reed, D. W., Underhill, E. W., Pomeroy, M. K., and Weber, N.,** Biosynthesis of acyllipids containing very long chain fatty acids in microspore-derived and zygotic embryos of *Brassica napus* L. cv. Reston, *Plant Physiol.,* 99, 1609, 1992.

236. **Wilen, R. W., Mandel, R. M., Pharis, R. P., Holbrook, L. A., and Moloney, M. M.,** Effects of abscisic acid and high osmoticum on storage protein gene expression in microspore embryos of *Brassica napus, Plant Physiol.,* 94, 875, 1990.

237. **Van Rooijen, G. J. H., Wilen, R. W., Holbrook, L. A., and Moloney, M. M.,** Regulation of accumulation of MRNAs encoding a 20 kDa oil-body protein in microspore-derived embryos of *Brassica napus, Can. J. Bot.,* 70, 503, 1992.

238. **Sangwan, R., Gauthier, D. A., Turpin, D. H., Pomeroy, M. R., and Plaxton, W. C.,** Pyruvate-kinase isozymes from zygotic and microspore-derived embryos of *Brassica napus, Planta,* 187, 198, 1992.

239. **Johnson-Flanagan, A., Huiwen, Z., Geng, X.-M., Brown, D. C. W., Nykifarutk, C. L., and Singh, J.,** Frost, abscisic acid and desiccation hasten embryo development in *Brassica napus* L., *Plant Physiol.,* 94, 700, 1992.

240. **Johnson-Flanagan, A. M. and Singh, J.,** Alteration of gene expression during the induction of freezing tolerance in *Brassica napus* suspension cultures, *Plant Physiol.,* 85, 699, 1987.

241. **Johnson-Flanagan, A. M., Barran, L. I., and Singh, J.,** L-Methionine transport during the induction of freezing hardiness by abscisic acid in *Brassica napus* cell suspension cultures, *J. Plant Physiol.,* 124, 309, 1986.

242. **Orr, W., Keller, W. A., and Singh, J.,** Induction of freezing tolerance in an embryogenic cell suspension culture of *Brassica napus* by abscisic acid at room temperature, *J. Plant Physiol.,* 126, 23, 1986.

243. **Orr, W., Johnson-Flanagan, A. M., Keller, W. A., and Singh, J.,** Induction of freezing tolerance in microspore-derived embryos of winter *Brassica napus, Plant Cell Rep.,* 8, 579, 1990.

244. **Johnson-Flanagan, A. M. and Singh, J.,** Degreening and its inhibition by stress in haploid embryos of *Brassica napus* cv. Topas and Jet Neuf, in *Proc. 8th Int. Rapeseed Congr.,* McGregor, D. I., Ed., 1991, 743.

245. **Anandarajah, K., Kott, L., Beversdorf, W. D., and McKersie, B. D.,** Induction of desiccation tolerance in microspore-derived embryos of *Brassica napus* L. by thermal stress, *Plant Sci.,* 77, 119, 1991.

246. **Brown, D. C. W., Singh, J., Takahata, Y., and Watson, E.,** Induction of desiccation tolerance in *Brassica napus* microspore-derived embryos, *In Vitro Cell Dev. Biol.,* 26, 71, 1990.

247. **Senaratna, T., Kott, L., Beversdorf, W. D., and McKersie, B. C.,** Desiccation of microspore-derived embryos of oilseed rape (*Brassica napus* L.), *Plant Cell Rep.,* 1, 342, 1991.

248. **Cloutier, S.,** *In Vitro* Selection for Freezing Tolerance Using *Brassica napus* Microspore Culture, M. Sci. thesis, University of Guelph, Ontario, 1990.

249. **Pauls, K.,** Cell culture techniques and canola improvement, in *Biotechnology of Plant Fats* and Oils, Rattray, J., Ed., American Oil Chemists' Society, Champaign, IL, 1990, 36.

250. **McClellan, D., Kott, L., Beversdorf, W., and Ellis, B. E.,** Glucosinolate metabolism in zygotic and microspore-derived embryos of *Brassica napus* L., *J. Plant Physiol.,* 141, 153, 1993.

251. **Bajaj, Y. P. S.,** Cryopreservation of pollen and pollen embryos, and the establishment of pollen banks, *Int. Rev. Cytol.,* 107, 397, 1987.

252. **Bajaj, Y. P. S.,** Regeneration of plants from pollen embryos of *Arachis, Brassica* and *Triticum* spp. cryopreserved for one year, *Curr. Sci.,* 52, 484, 1983.

253. **Chen, J. L. and Beversdorf, W. D.,** Cryopreservation of isolated microspores of spring rapeseed (*Brassica napus* L.) for *in vitro* embryo production, *Plant Cell Tissue Organ Cult.,* 31, 141, 1992.

254. **Charne, D. G., Pukacki, P., Kott, L. S., and Beversdorf, W. D.,** Embryogenesis following cryopreservation in isolated microspores of rapeseed (*Brassica napus* L.), *Plant Cell Rep.,* 7, 407, 1988.

255. **Chen, J. L.,** Evaluation of Microspore Culture in Germplasm Preservation, Lipid Biosynthesis and DNA Uptake Studies in Rapeseed, *Brassica napus* L., Ph.D. thesis, University of Guelph, Ontario, 1991.

256. **McKersie, B. D., Senaratna, T., Bowley, S. R., Brown, D. C. W., Krochko, J. E., and Bewley, J. D.,** Application of artificial seed technology in the production of hybrid alfalfa (*Medicago sativa* L.), *In Vitro Cell Dev. Biol.,* 25, 1183, 1989.

257. **Redenbaugh, K., Grian, D. P., Nichol, J. W., Kossier, M. E., Viss, P. R., and Walker, K. A.,** Somatic seeds: encapsulation of asexual plant embryos, *Biotechnology,* 4, 643, 1986.

258. **Datta, S. K. and Potrykus, I.,** Artificial seeds in barley encapsulation of microspore-derived embryos, *Theor. Appl. Genet.,* 77, 820, 1989.

259. **Mascarenhas, J. P.,** Gene activity during pollen development, *Annu. Rev. Plant Physiol.,* 41, 317, 1990.

Micropropagation

Kenneth L. Giles and Kenneth R. D. Friesen

Department of Horticulture Science, University of Saskatchewan, Saskatoon, Canada

Introduction

Of all the technologies and techniques that have influenced horticulture in the last 25 years, micropropagation has had the most significant impact in shaping and developing commercial horticulture. Unquestionably, significant contributions have also been made by developments in facilities and chemicals for disease and insect control. But micropropagation has shaped not only the practice of horticulture, but also to some degree the aspirations of the horticultural industry.

Micropropagation is the controlled *in vitro* propagation of plants under sterile conditions. The propagation is influenced by the inclusion of plant growth regulators in the culture medium. These growth regulators either overcome apical dominance within the plant shoots, leading to the development of axillary shoots, or induce adventitious shoot formation, again leading to the production of more shoot material. These shoots are separated and recultured on a similar medium in such a way that the process is repeated and shoot multiplication occurs. This process frequently results in exponential proliferation. Other objectives of micropropagation, in addition to multiplication, are elimination of pathogens, rapid introduction of new or novel genotypes, selection of somaclonal and induced variants, and preservation and long-term storage of germplasm.[33]

0-8493-8262-9/94/$0.00+$.50

The precise requirements for nutrients, growth regulators, and cultural and environmental conditions must be determined for each and every species and variety, as significant differences often exist.[13] This fine tuning of growth regulator and medium is necessary in order to optimize the system to provide consistent numbers of robust plantlets for soil establishment. The many publications relating to the 'bench level' observations regarding micropropagation are frequently too superficial and often misleading for practical large scale commercial propagation.[13]

Economics

There is increasing popularity and demand for micropropagated plantlets, the number in international trade having doubled during the last 10 years. Approximately 180 to 200 million plantlets are produced per year via tissue and cell culture worldwide.[33] Most laboratories produce herbaceous, mostly ornamental plants and rootstocks; fruit trees and small fruits, vegetables, and agricultural crops comprise the rest in descending order of magnitude.[13,33,42]

There has also been a significant increase in the number of commercial micropropagation laboratories around the world. There are currently 8 new laboratories being established in India, taking the total in that country to approximately 12. Much of their export material is going into Western Europe and finding a ready market because of the relatively low costs of labor in India. When compared to the costs of North American-produced material, the Indian material can frequently be half the cost of locally produced plantlets. Even with air freight, handling, and associated costs, the Indian-produced material is of significantly lower price, though in no way inferior in quality. Although the general trend is toward large, efficient laboratories, there is still a place for speciality work: virus clearing, breeding lines for F_1 hybrids, and test plants derived from genetic manipulation.[13,42]

To develop a commercial micropropagation laboratory, four diverse fields must be mastered: financing, knowledge of the market and access to it, research and development, and production management.[42] There is a significant number of popular articles suggesting that micropropagation, despite its sophisticated nature, can be carried out effectively and efficiently by small nurseries. These articles seem to suggest that a glorified kitchen approach to micropropagation is a worthwhile commercial enterprise and can be of advantage to small growers. In our opinion, this is not the case. Small nurseries have limited facilities, labor, and time. A well-controlled and efficiently run micropropagation lab demands continuous attention and significant capital outlays in facilities and equipment.[42] Almost more important than these outlays is knowledge, training, and experience in a variety of disciplines needed for the production of a wide spectrum of species and varieties. The movement from the laboratory to commercial production requires fine tuning and adjustment of procedures.[42] Such skill and the required information base do exist in large-scale commercial micropropagation labs where a team of experienced and well-trained personnel can be focused on problem solving and systems development.

It is important that small scale growers are not misled into thinking that micropropagation can be practiced in an efficient and effective manner on a small nursery scale. It would be more practical for them to contact centers of commercial micropropagation for advice and help in identifying varieties already in culture that could be available in a rapid, cost-efficient manner. Such inquiries would save them considerable capital investment and time commitment that might be more effectively used in nursery or greenhouse management. Most commercial micropropagation companies are willing to devote some research time to the development of new propagation systems for new species and cultivars introduced by growers. Such services are frequently costed out under separate contracts or occasionally can be built into the final cost of plantlets delivered.

The pricing of micropropagated plantlets depends on many factors, including labor content, multiplication rates *in vitro*, rooting percentages, and the intrinsic value of the plant material being propagated. Florkowski et al.[10] performed an analysis on the pricing of *in vitro* produced plantlets. The study analyzed factors that affect wholesale prices of selected micropropagated plantlets and compared them to price lists in 1987 from 14 tissue culture companies (10 from Florida, 2 in Texas, and 2 in California). The key factors identified were (1) the scale of operation as measured by total number of species and cultivars offered for sale, (2) the degree of specialization as measured by the number of cultivars per species sold, and (3) the location of the laboratory. It was hypothesized that the total number of plants offered for sale could lower the price per plantlet because of the economies of scale. The degree of specialization would also increase a firm's efficiency, thus lowering prices. Results, however, indicated that no strong inferences could be made regarding the impact of the scale of operation on wholesale prices of the selected micropropagated plantlets. The number of cultivars of the same species was consistently the most important variable influencing the price. As the degree of specialization increases, the price of the cultivar decreases. Consequently, a production lab must have a substantial part of its production in a few cultivars that are demanded in large quantities and at a reasonable price.[42] It was also noted that the location of the micropropagation laboratory did not have a direct impact on the price of a specific cultivar, there being considerable variability between laboratories in the same zone.

Micropropagated plants will always be more expensive than seeds but can have a higher intrinsic value in quality and quantity.[13] The value of the micropropagated plants also involves the time, space, materials, and labor saved by not using seed, especially in species with long dormancy or germination periods.

The costs of micropropagated plants break down to labor, 60 to 80%; growing space, 10 to 15%; and medium ingredients, 5%.[42] Of the total costs,

67% involve tissue culture manipulations,[42] 60% (40% of the total) of which involves *in vitro* rooting.[33] Some cost savings could be generated in this area through the use of *ex vitro* rooting and somatic embryos.[33]

Labor

Requirements for an efficient micropropagation system demand considerable labor, care, and attention, both at the commercial production level and in research and development. Well-trained and experienced labor to operate large-scale micropropagation facilities must usually be developed by each individual laboratory. Careful supervision of inexperienced staff is always required. Training with sterile technique can be achieved rapidly with most technicians. However, developing a comprehensive and well-defined background in a wide range of species and cultivar requirements can take considerable time. The development of efficient teams is aggravated in most circumstances by the demand for plants at particular times of year. The majority of nursery and greenhouse growers require nursery stock and outdoor planting material in the spring, such as *Gerbera,* which may make up a high percentage of many micropropagation labs' production. The demand profile results in peak labor demand over the winter periods, during which time large numbers of cultures are being produced, rooted, and hardened off for delivery. This results in a percentage of the staff being laid off for the summer and early fall months prior to recall for production runs the next year. Because of the necessity of a well-experienced workforce, this layoff jeopardizes the stability of any organization, because a certain number of the employees do not return the following year and their experience is therefore lost. Considerable efforts are made in most laboratories to smooth out the production throughout the year by looking at exporting into the other hemisphere in order to satisfy the demand there. Such arrangements can cause problems with the delivery of mother stock material for the initiation of cultures, but the phytosanitary requirements for *in vitro* propagated material shipped on agar is reasonable and well-interpreted almost everywhere.

Automation and Robotics

In some parts of Southeast Asia energy represents a very significant proportion of costs. However, in most Western European and North American laboratories the largest component in pricing is in labor costs. Labor can make up 60 to 80% of costs.[42] In response to labor costs, two lines of development are underway: automation and, as previously noted, development of laboratories in third world countries where wages are lower.[42]

Because of the substantial labor component in the production of micropropagated plantlets, considerable interest has been developed in mechanized, automated, or even robotic assistance of various stages of micropropagation. Such a mechanical system has the advantage of potentially lower contamination rates, since there would be less technician handling time and, if appropriate production rates can be achieved, considerable lowering of the labor force. Such steps are most appealing to companies in North America, Western Europe, Australia, and New Zealand, where labor costs are a considerable proportion of the final cost of the plant.

According to the Robotics Industries Association, a robot is a reprogrammable, multifunctional manipulator designed to move materials, parts, or specialized devices through variable programmed motions for the performance of a variety of tasks. It must have mechanical versatility and be auto-adaptable through sensory feedback with the goal of being self-controlled and self-monitoring. As such, robots must have a high degree of mental and physical agility.[27] This is in contrast to automation which is a mechanical device which performs a task with limited function and requires human assistance. While not as labor efficient, automated devices are productive and reduce the risk of contamination.[13]

Johnson[21] considered the requirements for any automated micropropagation system. Seven key factors were identified: (1) the sterility of plants must be maintained in the process; (2) media constituents must be determined and delivered at optimal levels; (3) there must be a stringent control of temperature, humidity, and light; (4) there must be a rapid analysis of the image of the plant structure, computing the optimal cutting strategy for each variety and driving appropriate tools; (5) the system must transfer correctly sized sterile pieces of tissue to the next stage, recording placement as well as tracking and timing clones through subsequent treatments; (6) viable plantlets must be transferred to final containers for hardening and other preplanting growth stages; and (7) the finished microplant must be delivered field-ready with appropriate supplements for planting in a prepared soil environment. This last stage is clearly not carried out *in vitro*, but is the logical end to the process. The one consideration that appears to be lacking in this analysis of requirements is that the overall unit must be affordable and efficient enough to be a cost-effective capital investment for any micropropagation company.

There has been greater research into automation in micropropagation than robotics. This is not unexpected given the greater complexity of robotic systems. The greatest advancement in this field has been in those stages of micropropagation that are the least costly. Systems are being developed and are available to reduce labor in media preparation and distribution.[1] Similar advances have yet to be made in automating or applying robots to the costly labor-intensive stages of *in vitro* multiplication, *in vitro* or *ex vitro* rooting, and greenhouse or field transfer. Shoot multiplication needs mechanization because it is repetitive and costly.[13] Systems developed at present are not yet flexible, quick, nor economical enough for general commercial applications.

Robotic applications to micropropagation requires further work in sterility and environment controls, visualization and identification of plant parts, and sensory capability of both grippers and end effectors.[13] Current research is progressing in the fields of computer visualization and artificial sensation.

Smith[40] described a robot work station used as a test bed for evaluating tools and techniques to facilitate the automation of micropropagation. A number of novel plant manipulation tools were developed and proved at a prototype level. These were mounted as tools on a cartesian geometry robot with 3 degrees of freedom plus wrist action. The robot was driven by stepper motors to perform the required cutting, harvesting, and planting movements. Overall robot control was provided by a sequential, predefined movement program which was updated according to the shape, position, and orientation of a particular microplant using information from a video camera. The video picture could either be displayed on a monitor and assessed by an operator or analyzed by a vision processing system. In the latter case, plant features were identified automatically and the robot, driven under computer control, performed the required task.

Tillett[45] gave results of successful and unsuccessful trials using a provision processing algorithm which could identify dissected parts of a plant and guide a robotic planter to plant them. The system used a corner finding technique to classify objects in the image and then calculated the location and orientation of the desired plant pieces. A similar computer vision algorithm was developed by McFarlane[28] for locating segments of stems of *Chrysanthemum* microplants in a container and selecting suitable segments for a harvesting robot to grasp. The system was tested with 32 images of eight microplant containers. The algorithm successfully identified 93% of the visible stems segments. In this case, the algorithm was designed to select stems through the bottom of the culture container, a problem which probably explains the number of errors identified.

In a study that acted as a model of automated micropropagation involving somatic embryos, He et al.[18] described a method of using seed for processing and analyzing images of living plants growing on cultivation medium. The procedure monitored seed germination and shoot growth status of the plants and described developing components such as cotyledons and leaves. The study demonstrated the potential for specially designed vision-controlled robots to locate suitable microplants and convey them to individual growing pots. Subsequent growth could then be monitored and controlled.

Computer Aided Management

As already stated, micropropagation is a complex, labor-intensive process serving a diverse, fluctuating market.[20] Commercial laboratories utilize two propagation systems: (1) the stock system, which is comprised of disease-free, true-to-type propagules frequently flushed with new material, and (2) the production cycle, which is comprised of propagules undergoing multiplication prior to planting and or sale. Added to each cycle are space utilization and contamination problems. This results in managerial nightmares balancing the ever-fluctuating demand with production and economic efficiency.

Computer-aided managerial programs have been developed both commercially and privately to assist the production manager with the complexities of large-scale commercial micropropagation. These production simulations must evaluate a large number of parameters and variables in order to provide accurate, relevant information. These include cultivar, multiplication rates, harvest rates, contamination ratios (both harvestable and disposal in both stock and production cycles), total number of production cycles, cycle number, rate new material enters the system, minimum stock requirements, stock-to-production ratio, demand, space, labor, medium used (and date it was prepared), container size, propagule density per container, and vessel location.[20]

The purpose of these simulations is to provide immediate information on clones and subculturing records, to allow easy retrieval of information on the status of orders, to keep track of the number of plants lost due to contamination, and to save time.[1] It is required, however, that the computer system used must be compatible with the organization and handling systems used in the laboratory and greenhouse.[1]

Contamination

Sources

Both fungal and bacterial contamination of *in vitro* cultures are everpresent threats in any large-scale micropropagation establishment. Contamination can occur as a result of a number of factors. The most common is the accidental introduction of bacteria or fungal spores into the culture as a result of handling during subculture. Although technical personnel are trained in sterile culture, the pressure of rapid, frequent subculture introduces the risk of contaminating organisms being present.

Leggatt et al.[26] reported that 31 microorganisms were isolated and characterized from 10 different cultivars growing in micropropagation. Yeast, *Corynebacterium* species, and *Pseudomonas* species were the most common isolates. Most of the isolates were capable of using a wide variety of carbon sources, and a number of them were resistant to antibiotics. Interestingly, reinoculation of 13 apparently disease-free plants resulted in only 4 displaying disease symptoms similar to those produced by the original microorganisms. In only one of these four cases did the microorganism reisolated appear similar to the organism used for inoculation, and even in that case, the reisolated microorganism was present with others, suggesting that disease-free plants carry a number of microorganisms while growing in micropropagation and that mixed cultures of microorganisms may be required to induce disease symptoms.

As yet, there are no industry-wide standards with respect to particle counts and hygiene standards.[13] Such standards must be economical and physically practical in a competitive, high-cost industry. The solution appears to lie in having a clean area concept, a strict code of hygiene, regular inspection and culling of cultures, and rapid, efficient identification of potentially contaminated material.[13]

Boxus and Terzi[5] reviewed the control of accidental contamination during mass propagation in laboratories producing 1 million plants per year. They identified four key factors in the spread of contaminating organisms: (1) insufficient sterilization of media, (2) insufficient flaming of the tools, (3) the use of contaminated alcohol in flaming, and (4) the spread of resistant bacterial lines. Contamination from insufficient autoclaving or inadequately sterilized media is now less common, particularly with well-controlled automatic autoclaves and careful monitoring of heat and pressure during autoclaving.

Once there is significant fungal contamination in a batch of cultures, a more serious form of contamination can occur with mites and thrips conveying fungal spores from one container to another, rapidly spreading fungal contamination.[13] Such effects can be almost epidemic when they occur, and there is absolutely nothing that can be done other than autoclaving the cultures and disposing of them. Indeed it is true of almost all contaminated cultures that the work and time involved in resterilizing and ridding cultures of contaminants is an expensive and almost worthless exercise. There are various antibiotics that can be added to the medium which will control bacterial contamination, but in almost all cases they will reduce the vigor of the plant cultures if used on a regular basis; coupled with this inconvenience is the obvious expense involved in using antibiotics in this way. Of the antibiotics tested, rifampicin, chloramphenicol, and neomycin were the most inhibitory to bacterial growth.[26]

Identification

In many cases the existence of contaminating organisms in micropropagated culture is all too obvious. Occasionally, however, there are trace contaminants which do not overrun the cultures but that are inhibitory to maximum culture performance. Boxus and Terzi[5] suggested the addition of peptone (265 mg/l) and yeast extract (88 mg/l) to media in order to reveal latent bacterial contamination. Under these conditions, examination of the clarity or turbidity of the condensation water at the agar surface facilitated the detection of contaminated flasks and their removal. The researchers also made the observation that the addition of kinetin (0.5 mg/l) to rooting media contaminated with bacteria allowed nearly normal development of young tree plantlets without inhibiting root induction.

DeBergh and Vanderschaeghe[7] stressed the importance of subtle symptoms as being the most useful aid to identify contaminated cultures when all other detection aids had failed. They pointed out that faint brown spots on the petioles of *Gerbera* or brown bracts at the base of tissue cultured *Maranta* were telltale signs of latent contamination.

Contamination for Benefit

Naryanan et al.[29] described a fungus isolated from the flower buds of the bottle brush *(Callistemon viminalis)* and tentatively identified it as a nonpathogenic strain of *Sphaeropsis tumefaciens* that showed growth-promoting activity in cultures. Inoculum prepared from *in vitro* cultures of the fungus inoculated under the roots of seedlings of *C. viminalis* and citrus spp. in axenic culture produced a two- to fivefold increase in the height and an increase in root mass and vegetative growth of plantlets. Plants inoculated with a wild-type fungus died within a few days. Anatomical studies of inoculated *Callistemon* seedlings showed that the fungus forms a sheath around the root similar to the Hartig net formed by ectomycorrhizas. Potted sunflower and sorghum plants inoculated at the first true leaf stage showed a 20% increase in overall height and a 30% increase in dry weight.

Such findings raised the interesting possibility of deliberate infection of cultures to either induce superior growth or to inhibit accidental infection by other more damaging microorganisms.

The use of *Agrobacterium rhizogenes* to induce root formation in difficult-to-root cultivars has also been suggested. Although an attractive idea, it presents potential problems of international distribution of a potentially harmful disease, in addition to the problems encountered with international phytosanitary regulations.

Somaclonal Variation

Somaclonal variation is any variation that can arise through the culture of plant cells, tissues, organs, and explants and their regenerated plants.[22] It is a major problem in micropropagation[6] where commercial application requires maintenance of high regeneration rates. This desired rate of multiplication is limited by declines in organogenic competence, genetic instability, and vitrification.[33]

In an appropriately controlled and maintained culture system, the development of genetic variance is rare. Somaclonal variance will occur if the cultures are allowed to produce excess callus from which adventitious and potentially genetically altered shoots may regenerate. Other sources of variation can be the length of time in culture, physiologic disorders due to mutagenic substances (e.g., growth regulators) or as a response to stress, directed developmental changes in chromosomes (epigenetic variation), explant tissue origin (meristematic vs. nonmeristematic), chimeral breakdown, and elimination of infectious agents.[6,16] These variations again emphasize the need for a well-tuned, well-defined culture protocol for maximum reliability.

Variance produced as a result of somaclonal variation tends to be mainly deleterious.[6] Grunewaldt and Dunemann,[17] however, feel that despite the generally deleterious nature of somaclonal variation, it does allow useful variants to be selected occasionally.

Schwenkel and Gruenwaldt[38] reported the *in vitro* culture of eight diploid genotypes of *Cyclamen persicum* that were grown in the greenhouse until flowering to give about 6000 plants. The frequency of altered plants ranged from 4.5 to 36.3% depending on the type of variation and the original genotype. Cytological, regeneration, and segregation analyses showed that most of the altered phenotypes were genetically controlled. Results indicate that the usefulness of selection of somaclonal variants from populations has significant value in the development of new ornamental varieties, but of course lacks any of the directed advantages of transformation or genetic manipulations. The numbers required for the successful selection of superior or improved genotypes are likely to be very significant, and the advantages and disadvantages of somaclonal selection methods must be carefully considered.

De Klerk et al.[6] suggest that an assay to measure the extent of somaclonal variation be developed. They suggest that it could be based on an assessment of the frequency of aberrant plants involving storing variation in the phenotype, chromosome number and size, electrophoretic protein and isoenzyme patterns, and DNA restriction fragments. As an alternative, an assay could be based on determining the spread of quantitative trait variation in the populations by calculating the coefficient of variation.

As has already been noted, commercial micropropagation must ensure genotypic invariability for each cultivar. When plantlets are exposed to plant growth hormones the stability of the cultures may be at risk. Endogenous concentrations of plant hormones in micropropagated plantlets are frequently uncertain. A series of profound physiological changes can overcome the cultures, including vitrification if the combined influence of external and internal growth regulators is excessive. These changes can significantly influence the ability of plantlets to root and establish at the end of their culture periods. This dictates the careful monitoring and control of plant growth media during long term culture.

Leaf variegation and chimeral cultivars also offer a problem in culture. Frequently, the chimera are either unstable or breakdown, and continuous selection pressures must be put upon the culture to maintain true-to-type plantlet production. The selection that has to be imposed on the cultures can involve considerable time and make such cultivars not economical for a commercial operation to attempt large scale propagation.

Vitrification

Once in culture, the plant tissue's reactions to growth regulator concentrations is liable to change with time. Thus, in long-term, large-scale propagation procedures concentrations of growth regulators frequently must be lowered during the course of *in vitro* culture in order to provide continuous shoot production.

Oversupply of growth regulators, particularly cytokinins, may lead to the condition called 'vitreous' tissue. This vitreous tissue is succulent, water-filled, and frequently ceases or dramatically slows growth. Vitrification results in disturbed mechanisms of stomatal movement, decreased lignification of the shoot, twisted chlorotic leaves, and uncontrolled water release.[33] In addition to overexposure to cytokinins, vitrification can be caused by water vapor saturation and increased ethylene and CO_2 concentrations.

Vitrification can be corrected in some species, but it is far wiser to cautiously control the growth regulator concentrations in order to avoid the condition in long-term propagation situations.

Long-Term Storage

Long-term storage of cultures offers a number of advantages that can be exploited with regard to the labor peaks mentioned previously. In off seasons, when demand is low, production can continue at a modest rate with cultures eventually being stored and reactivated closer to delivery times. Once in culture, plant material can be stored for protracted lengths of time between subcultures on media developed to promote slow growth or zero multiplication. Low temperatures and low lighting can also be used to slow down the growth and demand for subculture.

Much research has been conducted in the area of cryopreservation and is beyond the scope of this paper. More work in this area is required since every species is likely to have a somewhat different response to low temperature storage. Large scale storage of *in vitro* material is an attractive possibility once the conditions of dormancy and cold damage are better understood.

Cultural Innovations

Conventionally, plant tissue cultures have been grown in a variety of containers with little concern or quantitative measurement of internal atmospheric conditions. Light penetration has usually been a question of compromise, with many varieties of different species growing under the same light intensity and light regime. Increasingly, there is evidence to suggest that the useful optimization of these variables could have significant effects on productivity of the propagation systems.

Questions concerning the quality and constituents of the *in vitro* atmosphere have resulted in investigations into this variable. Figueira et al.[9] examined the influence of CO_2 *in vitro*. Using cocoa *(Theobroma cacao)*, which is normally recalcitrant in culture, nodal cuttings were induced to elongate and produce leaves under enriched CO_2 levels and higher light irradiances. This suggests that subculture of microcuttings under CO_2 enrichment could be the basis for a rapid system of micropropagation for cocoa.

Investigations into laboratory air pollutants have indicated that the use of bunsen burners and alcohol lamps in laminar flow hoods could be detrimental to the vigor of cultures. Combustion products of natural gases and ethanol (including ethylene) produced toxicity symptoms in *Prunus* cultures.[36] This indicates that the use of electronic heating units to sterilize dissecting tools in laminar flow hoods would be advantageous.

Investigation into optimizing culture media has gone beyond altering salt and growth regulator concentrations. Ghashghaie et al.[12] report interesting findings on the concentration of agar and water on rose plants cultured *in vitro*. The rose cultivar, Madame G. Delbard, was micropropagated *in vitro* at various agar concentrations from 0 to 15 g/l. Except for a liquid medium, where shoot proliferation was impaired by vitrification, fresh and dry weights and number of total shoots decreased linearly with increasing agar concentration. The number of usable elongated shoots was significantly higher at an agar concentration of 7.5 g/l. The authors explained this in terms of antagonistic actions of cytokinin and water on shoot elongation.

Continual flow, liquid nutrient medium has been promoted as an alternative to semi-solid agar.[47] This system uses microporous polypropylene membranes floating on a liquid nutrient. In addition to enhanced growth, *in vitro* plantlets on the microporous membrane are free from entanglement with the support matrix and readily available to mechanical handling. Such a growth system would allow mechanization for mass handling, separation, and transfer of plant tissue cultures.

Nutrient mist cultivation supplies a fine, ultrasonically produced mist which envelops plantlets resting on a metal grid. This method of culture has proven to be at least as productive as agar-based media for *Musa cordyline, Nephrolepis* spp., and *Solanum tuberosum*.[13a,25,46] The advantages of the nutrient mist technique are that less medium is necessary for growth; sterility is more rigorously contained by the provision of a positive-pressure environment for the plant cultures, and movement, mechanization, and eventual automation involving nutrient mist technologies are simplified when compared to agar-based culture systems.

Somatic Embryogenesis

Somatic embryogenesis is the vegetative generation of embryos from a single or a few cells. This generation may be directly from organized plant tissue, such as a leaf, or it may be indirectly formed through an intervening stage of callus. These embryos are similar to zygotic embryos in all regards except for their origin in vegetative tissue as opposed to fertilization. Somatic embryogenesis was first described by Steward[43] and Reinert[35] in carrot callus and suspension cultures. Since that time, the generation of somatic embryos has been reported in numerous species and cultivars.[2,3]

In principle, the implications and potential uses of somatic embryogenesis are enormous.[19] The development of this technology will permit rapid introduction of new cultivars and crops derived from genetic transformation. It also lends itself to automation and robotic manipulation[8,32] and holds the potential for the development of artificial seed.[11,30,34,39]

Somatic embryogenesis has the advantage that rapid batch culture can be used to produce thousands of propagules from a small amount of callus tissue. Little space is required as even a small bioreactor of a few liters in volume can produce enormous quantities of embryos.[32] The resultant decrease in labor, space, and materials translates to more affordable micropropagated material at a cost per unit approaching that of hybrid seed.[19]

At present, there are some significant obstacles to the widespread commercialization of this technology. Somaclonal variation in propagules originating from callus culture, as has been previously discussed, is a major problem. Somatic embryogenesis may not be any better (and possibly worse) than adventitious buds in regard to variability.[6,41] In species where mass plantings are common, as in forestry, variation between propagules is not a concern. In ornamental species such as cut flowers, however, such variability is unacceptable. Current production methods also have poor reproducibility as organogenesis in callus culture is not as good as conventional micropropagation using adventitious shoot induction.[32]

Other areas requiring further study are the physiology and genetics of somatic embryogenesis, long-term culture and plantlet development of somatic embryos, large-scale production technology of suspension culture and bioreactors, greenhouse and field establishment of plants derived from somatic embryogenesis, and subsequent evaluations of their performance.[8,11,44]

One problem currently being addressed is the appropriate delivery system to transfer *in vitro* somatic embryos to the greenhouse or field. As noted previously, a potential benefit of somatic embryogenesis is the development of artificial or synthetic seed. The delivery systems which have been proposed are encapsulation (either in sodium or calcium alginates),[11] fluid drilling (somatic embryos suspended in a nutrient gel),[15,37] desiccated coated embryos,[23,24] and desiccated uncoated embryos.[4] Encapsulated or coated embryos have the potential for the development of an artificial endosperm. This could incorporate nutrients, fungicides, growth regulators, and mycorrhizas within the matrix of the alginate.[11,19,31]

Procedures for the production of artificial seed have been developed for a variety of crops, including alfalfa *(Medicago sativa)*, celery *(Apium graveolens)* and *Brassica* spp.[33] Techniques important for ornamental and forest tree propagation are also being developed.[4,14,44] However, these procedures are not ready for commercialization at this time. Major obstacles have yet to be resolved which include the identification and application of suitable coating materials that are stable and nontoxic. Such coatings must have evenly distributed and appropriately sized pores to permit respiration and hydration yet prevent leakage of larger molecules.[39] The physiology and procedures for desiccation of somatic embryos still require further study as this condition appears to be necessary for the conversion from a developmental stage to one of germination.[4,44] Vegetative cloning of desired genotypes is the ultimate aim of this technology. Consequently, careful consideration should be made of the impact of large-scale artificial seed production upon genetic diversity in crops.[19]

Future Developments

The future of micropropagation promises improved availability of new and novel cultivars. The cost of plant material, especially those species traditionally vegetatively propagated, should decrease. The trend toward improved plant health and subsequent improvements to quality and yield will also benefit from improved technologies and expansion of micropropagation.

Further advances in micropropagation will evolve with developments in the following areas: (1) regulation of organ differentiation on a cellular level, (2) controlled bioreactors for somatic embryogenesis, (3) improved propagation techniques, (4) preservation and storage of germplasm, (5) improved shoot regeneration, (6) *ex vitro* rooting, (7) preconditioning plantlets for acclimatization, (8) prevention of a gradual decrease in regenerative potential, and (9) maintenance of genetic stability during long-term culture and somatic embryogenesis.[33]

Summary

It seems certain that over the next 10 years the nature of plant micropropagation is likely to change either to a highly mechanized or automated process or to one that involves techniques such as nutrient mist, growth of cultures on microporous polypropylene, or somatic seed production. Economic pressure is certain to occur for micropropagation companies in high labor cost areas of the world to either reduce the labor content of micropropagation or to develop close links with areas of low labor cost in order to subcontract developed technology to lower labor cost environments. In either case, micropropagation is now well-entrenched as a tool for the introduction of new varieties or disease free material into the horticultural industry worldwide.

References

1. **Aitken-Christie,** *Micropropagation,* Debergh, P. C. and Zimmerman, R. H., Eds., Kluwer Academic Publishers, Netherlands, 1991, 389.

2. **Ammirato, P. V.,** Embryogenesis, in *Handbook of Plant Cell Culture,* Vol. 1, Evans, D. A., Sharp, W. R., Ammirato, P. V., and Yamada, Y., Eds., Macmillan, New York, 1983, 82.

3. **Ammirato, P. V.,** Recent Progress in Somatic Embryogenesis, *IAPTC Newsletter,* 57, 2.

4. **Attree, S. M., Moore, D., Sawhney, V. K., and Fowke, L. C.,** Enhanced maturation and desiccation tolerance of white spruce [*Picea glauca* (Moench) Voss] somatic embryos: effects of a non-plasmolysing water stress and abscisic acid, *Ann. Bot. London,* 68 (6): 519, 1991.

5. **Boxus, P. and Terzi, J. M.,** Control of accidental contaminations during mass propagation, *Acta Horticult.,* 225, 189, 1988.

6. **De Klerk, G. J. and Bouman, H.,** Measurement of somaclonal variation, in *Integration of In Vitro Techniques in Ornamental Plant Breeding,* de Jong, J., Ed., Eucarpia, Wageningen, Netherlands, 1990, 56.

7. **DeBergh, P. C. and Vanderschaeghe, A. M.,** Some symptoms indicating the presence of bacterial contaminants of plant tissue cultures, *Acta Horticult.,* 225, 77, 1988.

8. **Durzan, D. J. and Durzan, P. E.,** Future technologies: model-reference control systems for the scale-up of embryogenesis and polyembryogenesis in cell suspension cultures, in *Micropropagation,* Debergh, P. C. and Zimmerman, R. H., Eds., Kluwer Academic Publishers, Netherlands, 1991, 389.

9. **Figueira, A., Whipkey, A., and Janick, J.,** Increased CO_2 and light promote *in vitro* shoot growth and development of *Theobroma cacao, J. Am. Soc. Horticult. Sci.,* 116, 585, 1991.

10. **Florkowski, W. J., Lindstrom, O. M., Robacker, C. D., and Simonton, H. R.,** Analysis of pricing plants grown in tissue culture, *HortScience,* 25, 1306, 1990.

11. **Fujii, J. A., Slade, D. T., Redenbaugh, K., and Walker, K.,** Artificial seeds for plant propagation, *Trends Biotechnol.,* 5, 335, 1987.

12. **Ghashghaie, J., Benckmann, F., and Saugier, B.,** Effects of agar concentration on water status and growth of rose plants cultured *in vitro, Physiol. Plant.,* 82, 73, 1991.

13. **Giles, K. L. and Morgan, W. M.,** Industrial scale plant micropropagation, *Trends Biotechnol.,* 5, 35, 1987.

13a. **Giles, K. L., Friesen, K., and Norakovski, D.,** The use of *in vitro* mist cultivation for the propagation and hardening-off of virus free potato plantlets, American Society for Horticulture Science, 89th Annual Meeting, *Hortscience,* 27, 698, 1992.

14. **Gingas, V. M. and Lineberger, R. D.,** Asexual embryogenesis and plant generation in quercus, *Plant Cell Tissue Organ Cult.,* 17, 191, 1989.

15. **Grey, D. J.,** Quiescence in monocotyledonous and dicotyledonous somatic embryos induced by desiccation, *HortScience,* 21, 820, 1986.

16. **Grier, T., Beck, A., and Preil, W.,** High uniformity of plants from cytogenetically variable embryogenic suspension cultures of poinsettia (*Euphorbia pulcherima* Wild ex Klotzsch), *Plant Cell Rep.,* 11, 150, 1992.

17. **Grunewaldt, J. and Dunemann, F.,** Variation and selection *in vitro,* in *Integration of In Vitro Techniques in Ornamental Plant Breeding,* de Jong, J., Ed., Eucarpia, Wageningen, Netherlands, 1990, 39.

18. **He, W. B., Beck, M. S., and Martin, W. J.,** Processing of living plant images for automatic selection and transfer, *Comput. Electron. Agric.,* 6, 107, 1991.

19. **Hobbelink, H.,** *Biotechnology and the Future of World Agriculture,* Zed Books, London, 1991.

20. **Humphries, S., Simonton, W., and Thai, C. N.,** Computer aided management of plant tissue culture production, *Comput. Electron. Agric.,* 6, 33, 1991.

21. **Johnson, B. J.,** Towards automation of crop plant tissue culture, *Aust. J. Biotechnol.,* 3, 278, 1989.

22. **Karp, A.,** On the current understanding of somaclonal variation, *Oxford Sur. Plant Mol. Cell Biol.,* 7, 1, 1991.

23. **Kitto, S. and Janick, J.,** Production of synthetic seed by encapsulating asexual embryos of carrot, *J. Am. Soc. Hortic. Sci.,* 110, 277, 1985a.

24. **Kitto, S. and Janick, J.,** Hardening treatments increase survival of synthetically-coated asexual embryos of carrot, *J. Am. Soc. Hortic. Sci.,* 110, 283, 1985b.

25. **Kurata, K., Ibaraki, Y., and Goto, E.,** System for micropropagation by nutrient mist supply, *Trans.* ASAE 34, 621, 1991.

26. **Leggatt, I. V., Waites, W. M., Leifert, C., and Nichols, J.,** Characterization of microorganisms isolated from plants during micropropagation, *Acta Hortic.,* 225, 93, 1988.

27. **Marchant, J. A. and Moncaster, M. E.,** Robotics and automation in agriculture, *Outlook Agric.,* 19, 221, 1990.

28. **McFarlane, N. J. B.,** A computer-vision algorithm for automatic guidance of microplant harvesting, *Compu. Electron. Agric.,* 6, 95, 1991.

29. **Naryanan, K. R., Johnson, D. K., and McMillan, R. T., Jr.,** Growth promoting effects of a non-pathogenic fungus, in *Proceed. Plant Growth Regulator Society of America,* Plant Growth Regulator Society of America, Ithaca, NY, 1989, 94.

30. **Ng, T. J.,** Use of tissue culture for micropropagation of vegetable crops, in *Tissue Culture as a Plant Production System for Horticultural Crops,* Zimmerman, R. H., Griesbach, R. J., Hammerschlag, F. A., and Lawson, R. H., Eds., Martinus Nijhoff, The Netherlands, 1986, 259.

31. **Pierik, R. L. M.,** *In Vitro Culture of Higher Plants,* Martinus Nijhoff, Boston, 1987, 222.

32. **Preil, W.,** Application of bioreactors in plant propagation, in *Micropropagation,* Debergh, P. C. and Zimmerman, R. H., Eds., Kluwer Academic Publishers, Netherlands, 1991, 425.

33. **Rauther, G.,** Current status and future prospects of large scale micropropagation in commercial plant production, *Food Biotechnol.,* 4, 445, 1990.

34. **Redenbough, K., Viss, P., Slade, D., and Fujii, J. A.,** Scale-up: artificial seeds, in *Plant Tissue and Cell Culture,* Green, C. E., Somers, D. A., Hackett, W. P., and Biesboer, D. D., Eds., Alan R. Liss, New York, 1987, 473.

35. **Reinert, J.,** Morphogenese und ihre Kontrolle an Gewebekulturen aus Carotten, *Naturwissen,* 45, 344, 1958.

36. **Righetti, B.,** Air pollutants from hydrocarbons and derivatives in micro-propagation laboratories: toxicity symptoms on tissue culture of the cherry rootstock colt (*Prunus avium* × *P. pseudocerasus*), *Plant Cell Rep.,* 9, 374, 1990.

37. **Schultheis, J. R., Cantliffe, D. J., and Chee, R.,** Optimizing a sweet potato [*Impomoea batatas* (L.) Lam.] root and plantlet formation by selection of proper embryo developmental stage and size, and gel type for fluidized sowing, *Plant Cell Rep.,* 9, 356, 1990.

38. **Schwenkel, H. G. and Gruenwaldt, J.,** Somaclonal variation in *cyclamen persicum* Mill. After *in vitro* mass propagation, in *Integration of In Vitro Techniques in Ornamental Plant Breeding,* de Jong, J., Ed., Eucarpia, Wageningen, Netherlands, 1990, 39.

39. **Smidsrod, O. and Skyak-Braek, G.,** *Trends Biotechnol.,* 8, 71, 1990.

40. **Smith, G. P.,** The application of robotics to horticultural micropropagation, Agrotique 89, in *Proceed. 2nd Int. Conf.*, Sagaspe, J. P. and Villeger, A., Eds., Marseilles, 1989, 179.

41. **Smith, M. K. and Drew, R. A.,** Current applications of tissue culture in plant propagation and improvement, *Aust. J. Plant Physiol.,* 17, 267, 1990.

42. **Standaert-de Metsenaere, R. E. A.,** Economic considerations, in *Micro-propagation,* Debergh, P. C. and Zimmerman, R. H., Eds., Kluwer Academic Publishers, Netherlands, 1991, 123.

43. **Steward, F. C., Mapes, M. O., and Mears, K.,** Growth and organized development of cultured cells. II. Organization in cultures grown from freely suspended cells, *Am. J. Bot.*, 45, 705, 1958.

44. **Tautorus, T. E., Fowke, L. C., and Dunstan, D. I.,** Somatic embryogenesis in conifers, *Can. J. Bot.*, 69, 1873, 1991.

45. **Tillett, R. D.,** Vision-guided planting of dissected microplants, *J. Agric. Eng. Res.*, 46, 197, 1990.

46. **Weathers, P. J., Cheetham, R. D., and Giles, K. L.,** Dramatic increases in shoot number and length for *Musa, Cordyline* and *Nephrolepis* using nutrient mists, *Acta Horticult.*, 230, 39, 1988.

47. **Young, R. E., Hale, A., Camper, N. D., Keese, R. J., and Adelberg, J. W.,** Approaching mechanization of plant micropropagation, *Trans. ASAE*, 34, 328, 1991.

Artificial Seeds: A Comparison of Desiccation Tolerance in Zygotic and Somatic Embryos

Bryan D. McKersie and Susan D. N. Van Acker

Crop Science Department, University of Guelph, Guelph, Ontario, Canada

Introduction

Artificial or synthetic seeds are functionally defined as somatic embryos engineered to be of use in commercial plant production.[1] Zygotic embryos derive from the sexual recombination of male and female gametes and, thus, in many species are not genetically identical to the parents. This genetic variability is minimal in the seeds of self-pollinated crops, such as barley and wheat, but can be quite substantial in open-pollinated species, such as alfalfa or hybrid varieties of corn or tomato. Seeds also harbor a number of pathogens which can be inadvertently spread from contaminated seed-production fields. Many important crops, primarily tropical crops or ornamental plants, are sterile and do not set viable seed which necessitates propagation by cuttings or other vegetative means and prevents convenient storage. Artificial seeds are seen by many as a means of overcoming these limitations and allowing the clonal propagation of large numbers of disease-free propagules. The technology has

0-8493-8262-9/94/$0.00+$.50

applications in a diverse range of plant species as an aid in the breeding of hybrid varieties, as a system of storage of valuable genetic resources, and as a means of large scale multiplication.

There have been several different concepts of synthetic seeds which basically can be classified on the basis of whether the somatic embryo is naked or encapsulated and whether hydrated or dry. The exact type of synthetic seed required depends on the specific application and specific commodity. Naked, hydrated somatic embryos can be used to propagate high value plants such as ornamentals, where the efficiency of somatic embryogenesis reduces costs below the current labor intensive methods of micropropagation. An encapsulated, hydrated somatic embryo can be handled mechanically and directly planted into peat plugs for greenhouse production. These propagules could then be transplanted to the field in the case of vegetables such as carrot or celery. The capsule surrounding the somatic embryo could be alginate or similar product which would maintain the hydration of the embryo allowing temporary storage and could contain nutrients and other growth supplements to enhance the germination and establishment of the seedling.[1]

Drying the somatic embryo, whether naked or subsequently encapsulated, allows longer term storage of the propagule, and the synthetic seed then becomes a true analog of conventional seed. These dry somatic embryos could be used for long-term germplasm conservation or simply for bulking up propagules prior to greenhouse production. Dry somatic embryos could be easily distributed from a central tissue culture facility to a number of distant greenhouse production facilities.

Desiccation of somatic embryos may also have the added advantage of acting as a developmental cue in the maturation of somatic embryos. The seedlings from dry somatic embryos are, at present, less vigorous than seedlings from normal seeds,[2] and as a result, direct field plantings of artificial seeds is impractical with the present technology. This is a consequence of our poor understanding of the maturation process in seeds and therefore our inability to reproduce this maturation process *in vitro*. The inadequate deposition of storage reserves in somatic embryos is one example of improper embryo maturation. The processes involved in the embryo's preparation for water loss and the developmental consequences of desiccation are also poorly understood in developing seeds and therefore not readily imitated in artificial seeds. The following review attempts to briefly summarize our present understanding of these processes in zygotic embryos and our early efforts to transfer this information to the developing somatic embryo to produce artificial seeds.

Somatic and Zygotic Embryo Development

One of the main objectives of artificial seed production is to produce a propagule that is genetically, developmentally, and morphologically as close as possible to the seed of the species from which it has been derived. Thus, we will first review the similarities and differences between somatic and zygotic embryos (seeds) during development in order to appreciate the factors which induce desiccation tolerance, as well as the resultant variation in the level of desiccation tolerance.

Zygotic embryo formation begins with the double fertilization of the egg nucleus and the polar nuclei. Somatic embryos are formed by the differentiation of dedifferentiated somatic cells and are theoretically a clone of the donor plant.[1] In angiosperms, growth of both somatic and zygotic embryos begins with organized cell division in which polarity of the embryo is established.

The most obvious difference between somatic and zygotic embryo development is that zygotic embryos are nurtured by the mother plant, whereas somatic embryos are dependent on culture medium. Zygotic embryos are surrounded by an embryo sac and integuments which connect the embryo to the rest of the ovary. The integuments receive nutrients from the vegetative parts of the plant. The endosperm, in turn, grows by absorbing nutrients from the integuments, and in many species the embryo acquires nutrients from the endosperm.[3] The structures surrounding the embryo not only provide nutrients to the embryo but also serve protective and structural functions.[1] They act to control gas exchange as well as water and osmotic changes in the embryo.[4] As the seed matures, the integuments become the seed coat, protecting the growing seed from attack by diseases and mechanical stress.

Somatic embryos are grown on a culture medium consisting of carbon and nitrogen sources, minerals, and growth regulators.[5] Somatic embryos are induced from somatic cells which are developmentally plastic or competent. Like zygotic embryos, somatic embryo development is controlled by the physiochemical environment determined by cell culture medium.[6] Both somatic and zygotic embryos pass through similar morphological stages of development, including globular, heart, torpedo, and cotyledonary stages.[1,7,8] At the biochemical level there are basic similarities and distinctions between somatic and zygotic embryos, e.g., the storage reserves in alfalfa embryos. The storage proteins in alfalfa embryos were characterized initially by Stuart[9,10] and in detail by Krochko and Bewley.[11–14] In the mature seed, 30% of the total protein is an 11S storage protein, medicagin, which is a high-salt soluble globulin; 10% is a 7S storage protein, alfin, which is a low-salt, water-soluble globulin; and 20% is a 2S low molecular weight, sulfur-rich albumin.[11–13] These storage proteins are deposited not only in cotyledons but also the seed axis and accumulate to levels of approximately 25% of the seed dry weight. During germination, medicagin is rapidly hydrolyzed to amino acids and disappears within 48 h of imbibition. Storage protein synthesis in somatic embryos is qualitatively similar to that of zygotic embryos in alfalfa,[14,15] and similar results have been observed in Norway Spruce;[16] however, somatic embryos accumulate only about 10% of the storage protein levels measured in zygotic embryos. This is apparently controlled by the nutrient composition of the maturation medium, and glutamine seems to play an essential role.[15]

Seeds of the Trifolieae family, which includes alfalfa, retain an endosperm tissue inside the seed coat that contributes storage carbohydrates to the germinating embryo from cell wall hemicelluloses.[17,18] Starch and lipid reserves in the cotyledons are very low in the mature alfalfa seed. About 9% of the alfalfa seed's dry weight is a galactomannan, containing approximately 45% galactose, in the endosperm cell wall.[17,18] The outer cell layer of the endosperm, the aleurone, is devoid of galactomannan, but during germination it secretes hydrolases such as α-galactosidase, β-mannosidase, and β-mannase. Within 2 to 3 d of imbibition, these hydrolases degrade the galactomannan into galactose and mannose which are taken up by the cotyledons and metabolized into sucrose. In guar, starch is synthesized in the cotyledons as a temporary reserve,[19] and this probably also occurs in alfalfa because β-amylase activity increases by day 6 after imbibition.[20] In contrast, somatic embryos of alfalfa contain 10 to 15% starch on a dry weight basis,[21,22] and since they are obviously devoid of an endosperm, somatic embryos lack the galactomannan reserves of the seed. The pattern of carbohydrate utilization during the germination of somatic and zygotic embryos must be dramatically different, and this may contribute to the comparatively low vigor of dried somatic embryos.

Desiccation and Embryo Development

The development of desiccation tolerance and subsequent desiccation involves changes at the cellular and subcellular level in the embryo. The significance of these processes in the development of seeds is somewhat controversial.[4,5,23–27] The development of desiccation tolerance is a genetically controlled process which may be initiated by exogenous factors secreted from the maternal tissue or endogenous factors from the embryo itself. Desiccation may be just one of several temporally discrete developmental programs occurring within the embryo, each program being initiated by a separate factor that accumulates and induces the expression of a particular set of genes.[26] However, desiccation is not an obligate part of embryo development. Although in *Ricinus communis* desiccation terminated development and initiated processes associated with germination,[25] in several other species immature embryos placed on culture media are able to germinate without a period of desiccation.[3,28] Walbot[4] proposed that desiccation or developmental arrest involves three tandemly regulated genetic programs, including development, preparation for desiccation, and preparation for germination. Normally, seeds would progress from differentiation to germination except for the presence of physical and environmental restraints on the embryo. Implicit in her theory is the assumption that desiccation leading to developmental arrest is a bypass of the ordinary developmental program. The embryo is programmed to germinate while still immature, but the genes associated with germination are transcribed only when the embryo is removed from the plant.[3,4] Desiccation and subsequent imbibition are presumably critical to the removal of any germination inhibitors which prevent transcription or translation of these genes.

Somatic embryos normally germinate precociously without a desiccation phase unless efforts are made to prevent it, e.g., transfer of the developing embryo to a special maturation culture medium that mimics the zygotic embryo environment.[5] Details of the composition of the culture media necessary to promote embryo maturation in alfalfa *(Medicago sativa)* in our laboratory are described elsewhere.[5] Although the biochemical and physiological ramifications of desiccation to somatic embryo development may not be predictable, these efforts are predicated on the assumption that somatic embryo quality, and therefore artificial seed vigor, would be optimal if somatic embryo development emulated zygotic embryo development. From the above brief discussion of desiccation in seeds, the basic control of the expression of desiccation tolerance appears to be genetic, with regulation by endogenous factors and the embryo environment. A critical role for abscisic acid (ABA) has been shown in a mutant of *Arabidopsis thaliana* that is unable to synthesize ABA; the seeds usually do not desiccate on the plant and often exhibit vivipary.[29] Several studies propose that ABA plays a major role in the induction of desiccation tolerance in somatic embryos,[30–33] and one can speculate that ABA may induce the expression of a specific set of genes whose products facilitate the survival and subsequent desiccation.[34]

Variation in the Level of Desiccation Tolerance

Desiccation tolerance is not an absolute, unequivocal trait, but is expressed to varying degrees in different situations. In the context of embryo development, desiccation should be defined as the loss of cytoplasmic water to less than 0.20 g water per dry weight, a situation which is in equilibrium with ambient relative humidity. Similarly, desiccation tolerance is not simply the ability to withstand momentary water loss, but should be defined as the ability to regrow after a prolonged period of time in the desiccated state, usually several weeks. The inconsistent use of these terms has led to some confusion because varying degrees of desiccation tolerance have been induced in somatic embryos.

There are two methods used to dry somatic embryos that should be distinguished. The rate of water loss from the embryo or any tissue is dependent on the gradient in vapor pressure between the tissue and the atmosphere. All drying procedures involve transferring embryos from tissue culture media to atmospheres of defined relative humidity. In many studies, the humidity is kept constant in the range of 70 to 80%, and the embryos slowly equilibrate over time to this humidity. In other studies, the embryos are sequentially transferred from high humidity to progressively lower humidity with at least one day equilibrium at each humidity. Humidity is readily controlled on a small scale

Table 1 *Examples of the saturated salt solutions used to generate specific relative vapor pressures (relative humidities) for drying somatic embryos*

Salt	Relative vapor pressure
K_2SO_4	0.97
Na_2CO_3	0.87
NaCl	0.76
NH_4NO_3	0.63
$Ca(NO_3)_2$	0.51
K_2CO_3	0.43
CH_3COOK	0.20
LiCl	0.08

Note: The embryos are either placed in one atmosphere for a specific period of time or transferred sequentially through progressively lower humidities.

in desiccators containing saturated salt solutions. Examples of the saturated salt solutions used in our laboratory are given in Table 1, along with the theoretical relative humidity that each will generate. It should also be noted that the rate of drying will also depend on the size of the desiccator, the volumes of the salt solutions and the atmosphere, number of embryos being dried, and other factors which would influence the vapor pressure gradient.

In the first attempts to desiccate somatic embryos as artificial seeds, carrot embryos were encapsulated in a polyox coating prior to drying. Encapsulation improved embryo survival which may have been the result of reduced water loss but this was not recorded. Nonetheless, the embryos did not survive beyond 24 h.[35]

The somatic embryos of Sitka spruce — *Picea sitchensis* (Bong) Carr. — and interior spruce — a hybrid of *Picea glauca* (Moench) Voss. and *Picea engelmannii* Parry — survived after partial drying in a 95% relative humidity (RH) environment, but did not survive at 81% RH.[36,37] In another study, white spruce *(Picea glauca)* somatic embryos were dried through a sequence of relative humidities and equilibrated at the final humidity for at least 7 d. Survival declined as the embryos were dried below 81% RH, but 10% of the embryos survived drying at 43% RH.[38,39] Similarly, celery *(Apium graveolens)* somatic embryos survived equilibration at 90 and 70% RH but not at 50% RH.[40] Orchardgrass somatic embryos survived desiccation at a constant 70% RH to a water content of 13%; however, regeneration ability rapidly declined with even one week of storage.[41]

In alfalfa, somatic embryos were dried slowly through a sequence of relative humidities over a 7-d period or rapidly in 1 d at a room humidity of 30 to 40% RH. In both instances, the final moisture content was 15%, and after three weeks in the dried state over 90% of the visually selected, high quality embryos survived.[2,30,31] These embryos have been stored for one year at this moisture content with only a minimal loss in viability.

There are numerous factors which may have contributed to the differences in survival after desiccation among these various reports, but from our experience with alfalfa somatic embryos, the method of drying is not as critical as the tissue culture procedures used to mature the somatic embryo and induce the expression of desiccation tolerance.

Induction of Desiccation Tolerance

Numerous methods have been used to induce desiccation tolerance in somatic embryos, including exposure to exogenous ABA and treatment with sublethal stress.[2] The most common treatment is the inclusion of ABA in the maturation medium.[2,30,31] The critical importance of the induction phase to the subsequent survival of desiccation is shown in Figure 1. Somatic embryos at the late torpedo-early cotyledonary stage of development progressively lost the ability to germinate as they were dried sequentially through 76, 63, and 51% RH. In contrast, somatic embryos at the same stage of development were treated with ABA for 2 weeks and survived drying to approximately 20% moisture with over 90% germination.

ABA is present during seed formation and rises to a maximum level of 1 to 10 μM at mid-development, but may peak more than once during seed development and maturation, depending on the species.[42–44] In several species, changes in ABA levels have been associated with seed desiccation.[45–47] As well, ABA inhibits precocious germination of zygotic embryos[48,49] and of somatic embryos.[30,32] Thus, ABA has been used to induce tolerance to some degree of water loss in somatic embryos of zonal geranium *(Pelargonium x hortorum)*,[33] alfalfa (*Medicago sativa* L.),[30] celery,[50] white spruce,[38,39] and sitka spruce.[37]

Another means of inducing tolerance involves the application of a sublethal stress. Thermal stress and cold treatment have induced tolerance to dehydration in several genotypes of microspore-derived embryos of *Brassica napus*.[51] In alfalfa, sublethal stresses such as nutrient deprivation, cold stress, thermal treatment, and water stress all induce tolerance to desiccation.[30] Partial water stress can be applied in two ways. Senaratna et al.[30] simply removed the seal from the Petri plate containing medium and embryos, letting the moisture stress occur over 23 d. A more controlled method was that employed by Attree et al.[38] who used polyethylene glycol (PEG-4000) to simulate water stress in somatic embryos of white spruce. The mechanism of the ABA and stress treatments is not clear. ABA may confer tolerance by inducing a minor osmotic stress[52] or, alternatively, the sublethal stress may promote the endogenous accumulation of ABA in the somatic embryo.[53]

Sucrose and other sugars or sugar alcohols are osmotically active and hence have been assumed to play a role in desiccation tolerance. The desiccation

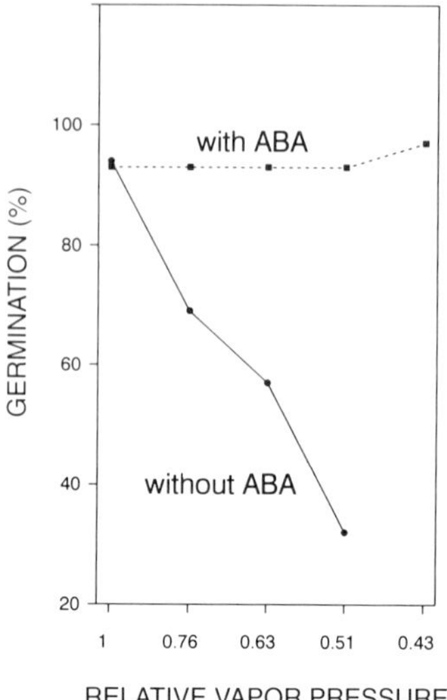

Figure 1. *Desiccation tolerance in somatic embryos of* Medicago sativa *line RL34 which were treated with abscisic acid (ABA) at the early cotyledonary stage of development and sequentially dried through atmospheres of progressively reduced relative vapor pressure.*

tolerance of alfalfa somatic embryos was increased simply by the inclusion of 6% sucrose in the maturation medium.[32] Several species including soybean (*Glycine max* [L.] Merr.), orchardgrass, and grape have survived water loss without any planned inductive treatment.[41,54] In these instances, the composition of the medium was conducive to maturation of the somatic embryo without any specific inductive signal.

Timing of the inductive treatment and the rate of desiccation are important. Alfalfa somatic embryos respond best when placed on ABA-containing medium at the torpedo-cotyledonary stage of development. If the embryos are introduced to the ABA prior to or after this stage, regrowth after desiccation declines.[31] In seeds of castor bean, slow drying results in higher rates of desiccation tolerance as opposed to rapid drying.[25] This is also true for *Brassica napus* microspore-derived embryos[51] and somatic embryos of alfalfa[5,30] and zonal geranium.[33]

Cellular Changes During Desiccation

Cellular changes that occur during maturation and desiccation involve structural alterations and fluctuations in the type of water binding,[27] as well as shifts in molecular organization and gene expression.[55–58] Many effects seem correlated with the acquisition of desiccation tolerance, whereas other effects seem to be a general response to water loss. It is beyond the scope of this review to comprehensively detail the cellular and molecular changes occurring during desiccation or the induction of desiccation tolerance in seeds which have been extensively reviewed elsewhere.[3,26,27,34,59] Instead, this section will focus only on recent results from our laboratory on desiccation of alfalfa somatic embryos.

Fluctuations in Water Binding

The water content of an embryo or similar plant tissue is dependent on the relative humidity of the atmosphere and on the ability of the embryo to bind water, termed *sorption.* Water sorption in plant embryos and seeds can be defined mathematically using the D'Arcy/Watt equation:[60]

$$W = \frac{KK'(p/p_o)}{1+K(p/p_o)} + c(p/p_o) + \frac{kk'(p/p_o)}{1-k(p/p_o)}$$

where W = the amount of water sorbed per gram of tissue; p/p_0 = relative vapor pressure; K = the attraction of strong binding sites for water; K' = the number of strong binding sites; c = the number and strength of weakly binding sites; k = water activity of multimolecular water; and k' = the number of multimolecular sorption sites.

This model takes into account three types of interstitial water binding: strongly bound, weakly bound, and multimolecular water (Figure 2). Vertucci and Leopold[61–63] have applied this model to desiccating orthodox and recalcitrant species. The fronds of *Polypodium polypodioides* have more strong binding sites for water, as well as a higher affinity for water, compared to the more sensitive species *P. vulgare*. Soybean embryos which have been germinated for 48 h lose their desiccation tolerance and, if subsequently dehydrated, show a decrease in the number of strong binding sites.[61,62] This is consistent with a previous observation that desiccation causes a loss of cellular elements such as phospholipids and intrinsic membrane proteins that normally bind water tightly.[64] Differences in the affinity for water binding between desiccation-intolerant and -tolerant seeds is determined by examining water content as a function of the relative vapor pressure (p/p_0) at which they are equilibrated; the resulting sorption isotherms differ in shape (Figure 3). Desiccation tolerant embryos show a reverse sigmoidal shape with convex, linear, and concave regions as shown for stage VIII alfalfa zygotic embryos, whereas the intolerant embryos have a hyperbolic shape with no convex region as shown for stage V embryos (Figure 3).

In experiments with embryos of alfalfa, we have determined that the high affinity binding sites differ between zygotic and somatic embryos, as well as between desiccation intolerant and tolerant stages of development. Desiccation-tolerant zygotic embryos (stages VII to VIII) were removed from the developing seed and produced sorption isotherms containing the typical three

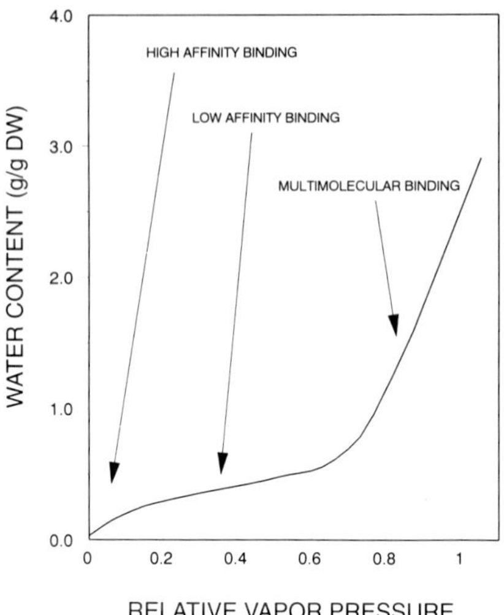

Figure 2. *Model of moisture isotherms indicating equilibrium water content of embryo related to atmospheric relative vapor pressure.*

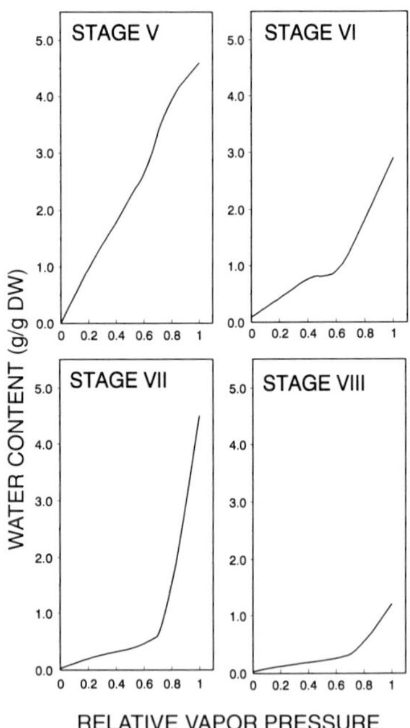

Figure 3. *Moisture isotherms of developing zygotic embryos of* Medicago sativa *cv. Excalibur. Embryo stages are those defined in Xu et al.*[7] *Stages V and VI are desiccation-sensitive, and stages VII and VIIIB are increasingly desiccation-tolerant.*

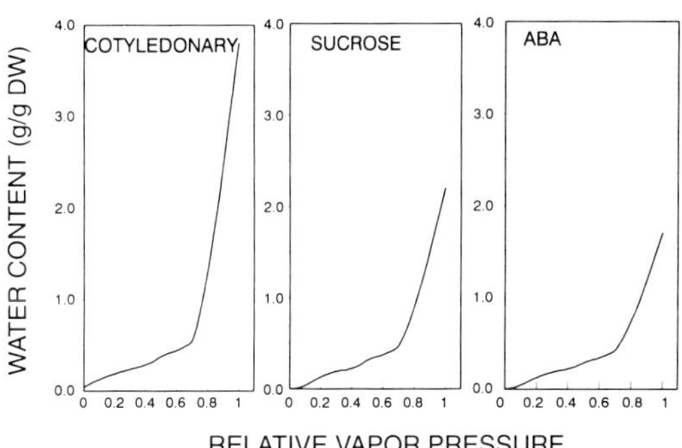

Figure 4. *Moisture isotherms of developing somatic embryos of* Medicago sativa *line RL34. Cotyledonary stage embryos are sensitive to desiccation while embryos that have been matured on ABA or sucrose are tolerant to desiccation.*

types of water binding: strong binding, weak binding, and multimolecular water (Figure 3). The region of strong binding is not present in stages V and VI, at which times the embryos are intolerant to desiccation. In contrast, mature, desiccation-tolerant somatic embryos do not follow the typical isotherms in that the shape of the strong binding region is concave instead of convex (Figure 4). Water loss from mature somatic embryos is complete before they are placed in 0% RH. It is possible that the somatic embryos do not contain strong binding sites for water or that these sites are lost as desiccation proceeds beyond a critical point. In seeds of other species studied, the strong binding sites are necessary for desiccation tolerance;[28,61–63] however, somatic embryos seem to be qualitatively different in the ways they bind water at low relative vapor pressure (RVP). As previously mentioned, somatic embryos of alfalfa differ from zygotic embryos in storage composition which primarily consists of starch in somatic embryos as compared to protein in zygotic embryos.[22] The zygotic embryos in Figure 3 are devoid of the endosperm and testa and therefore lack the galactomannan which also influences water sorption.[65] The differences in the storage components may result in qualitative or quantitative differences in the overall cellular components that bind water.

Developing zygotic and somatic embryos of alfalfa reduce the relative number of strong binding sites as desiccation tolerance increases (Figure 5). One explanation for this is that strong binding sites are not necessary for survival during desiccation. Another, conceivably more valid explanation is that of the water replacement hypothesis (WRH) proposed by Crowe.[66–68] The WRH maintains that the replacement of structural water by polyhydroxyls such as sugars and sugar alcohols will maintain the lamellar structure of cell membranes at low tissue water contents. As plant membranes lose water, it is believed that the phospholipid bilayer changes from a liquid crystalline phase to either a gel[69] or a hexagonal phase[70] at a distinct moisture content. The presence of sugars, such as sucrose in plants or trehalose in yeasts, reduces the

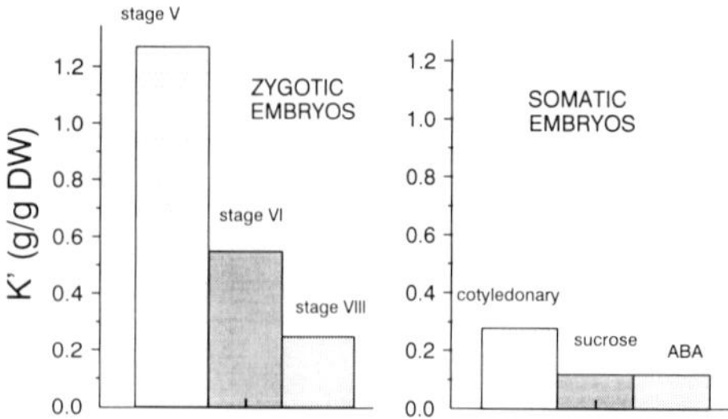

Figure 5. D'Arcy/Watt parameter K' which is representative of the number of strong binding sites in Medicago sativa *zygotic and somatic embryos. Stages representing desiccation-intolerant and -tolerant embryos are shown.*

threshold moisture content at which this transition occurs and therefore maintains the membrane's lamellar structure.[66,67]

Values for the D'Arcy/Watt water binding constants are determined on the basis of water content. If the water binding sites are occupied by sucrose in dehydrating embryos of alfalfa, the apparent number of strong water-binding sites as detected by this technique would decrease, even if the actual number of binding sites on the proteins and phospholipids remained the same.

Sugar Content Changes

There are several low molecular weight compounds that accumulate during the development of desiccation tolerance or at the onset of desiccation.[27,71] Bianchi et al.[72] have established that in *Boea hygroscopia* leaves, sucrose is the only sugar present in dry leaves. Glucose and fructose are present in fresh leaves but are detectable in only trace amounts in dry leaves, whereas sucrose increases over sixfold during drying. Raffinose and stachyose as well as sucrose accumulate in desiccating seeds of several species.[71] In model membranes, trehalose, a disaccharide, protects cell membranes by intercalating between phospholipids and proteins.[67,69] Induced hydrogen bonding would help to maintain the stability of the lipid bilayer on dehydration. Thus, it appears that desiccation-tolerant cells preferentially accumulate disaccharides and/or oligosaccharides and exclude monosaccharides.

Another means of protection by solutes is glass formation. Using sugars present in corn embryos as a basis, Koster[73] hypothesized that as desiccation proceeded the protoplasm would become quite concentrated, viscosity would increase, and solutes might potentially crystallize. As viscosity reaches a point at which no more water is able to diffuse, the solution should assume the mechanical qualities of a plastic solid or glass. Glass formation may be important in desiccation tolerance as a means of filling space in the cell, thereby helping to prevent collapse during water loss. When the sugar mix of raffinose and sucrose commonly found in desiccation-tolerant corn embryo axes was slowly desiccated, glass formation occurred at temperatures greater than 0°C,

Table 2 *Sugar composition (gram per gram dry wt) of* Medicago sativa *L. seeds at desiccation-intolerant (24 d after pollination) and -tolerant (36 d after pollination) stages and somatic embryos at desiccation-intolerant (without abscisic acid) and -tolerant (with abscisic acid) stages*

	Seeds		**Somatic embryos**	
Sugar	**24 d**	**36 d**	**−ABA**	**+ABA**
Stachyose	2	2	nd	nd
Sucrose	13	11	23	24
Glucose	25	7	37	3
Fructose	15	9	18	9
Galactose	14	3	0.5	0.6

but a sugar mix representing desiccation-intolerant axes composed of sucrose and monosaccharides formed a glass phase only at temperatures below 0°C. Raffinose may have been the key to glass formation and, therefore, desiccation tolerance.

Sugar contents of somatic embryos of alfalfa have been examined (Table 2). In contrast to seeds, the acquisition of desiccation tolerance induced by the inclusion of ABA in the maturation medium resulted in no net accumulation of sugars in the somatic embryos, but there was a distinct decline in the levels of the two reducing sugars, glucose and fructose.

Oxidative Stress

Cellular membranes are clearly one of the primary sites of desiccation injury, but different types of injury occur. One type of lesion involves the physical rupture of the membrane due to its inability to withstand the contraction and expansion associated with water loss and imbibition.[74] In addition, there are structural changes in the phase properties of the lipid as water is removed from the bilayer and its hydrophobic-hydrophilic interactions are perturbed. These changes in the lipid bilayer are thermodynamically reversible on rehydration, although membrane reorganization may not be.

After desiccation, membrane permeability properties are irreversibly altered. This is associated with irreversible changes in membrane structure. In lethally desiccated soybean axes, the microsomal membrane fraction had dramatically altered properties: the liquid crystalline to gel-phase transition temperature increased 38°C; lipid microviscosity increased twofold; the free fatty acid to phospholipid ratio increased tenfold; and the phospholipid to sterol ratio decreased by 50%.[64] All of these changes in the physical and chemical properties of cellular membranes can be simulated *in vitro* when the isolated membranes from healthy tissue are treated with activated oxygen (oxygen-free radicals).[75] The loss of desiccation tolerance in germinating soybean seeds is associated with the loss of lipid soluble antioxidants.[76] This observation suggests

Table 3 *α-Tocopherol content of* Medicago sativa *L. seeds at desiccation-intolerant (24 d after pollination) and -tolerant (36 d after pollination) stages and somatic embryos at desiccation-intolerant (without abscisic acid) and -tolerant (with abscisic acid) stages*

Tocopherol content relative to	Seeds		Somatic embryos	
	24 d	36 d	–ABA	+ABA
Dry weight (nmol/g)	108	337	443	785
Phospholipid (mmol/mol)	5.9	14.7	22.7	35.8

that desiccation promotes the formation of activated oxygen species which degrade membrane lipids, irreversibly change membrane lipid phase properties, and render the cell susceptible to rupture or metabolic dysfunction upon rehydration.

Oxidative stress also occurs in seeds during storage.[77] Therefore, one component of desiccation tolerance of somatic embryos would be an increased titer of lipid soluble antioxidants which would not only protect cell membranes during the initial drying process but also contribute to their longevity in the dried state. Very little analytical evidence exists to support this assumption, but in alfalfa the level of α-tocopherol (vitamin E) increased when the somatic embryos were treated with ABA to induce desiccation tolerance (Table 3).

It is also significant to note that the treatment of alfalfa somatic embryos with ABA at the late torpedo/early cotyledonary stage of development not only induces the embryo to express desiccation tolerance but also causes the embryo to degreen.[31] The excitation of chlorophyll by light is a major source of activated oxygen species in photosynthetic tissue.[78] Therefore, the loss of chlorophyll from these desiccation tolerant embryos is probably a major protective mechanism that reduces the potential for activated oxygen production, thereby contributing to the longevity of the embryo in the dry state.

Gene Expression

Gene expression during desiccation is affected at both the transcriptional and translational levels. Dehydrated rice callus displayed increasing levels of the transcript rab16A as drying time was lengthened. The transcript was also present in mature rice seed and prior to desiccation in callus that was treated with ABA.[79] The treated callus was tolerant to desiccation, yet the untreated callus was not. In corn, a specific mRNA increased in leaves during dehydration.[80] The protein encoded by this mRNA has been found to be high in glycine and contains repetitive amino acid sequences. The protein encoded by rab16A is also glycine-rich with repetitive amino acid sequences.[79] Leaves of the resurrection plant *Craterostigma plantaginium* also increased specific mRNAs in response to desiccation, and callus made desiccation tolerant by the application of ABA exhibited the same mRNAs. Not all the mRNAs were induced at the same time in leaves or treated callus, indicating that a gene network was involved, each gene or gene family with its own function, activated in succession.[56]

Other studies by Blackman et al.[81] and Lane[55] have dealt with specific soluble proteins that accumulate prior to desiccation of soybean and wheat seeds, respectively. Based on the many instances of the presence of ABA coincident with desiccation tolerance, these and several other authors have postulated that there is a mechanism in the embryo that responds to ABA to induce desiccation tolerance.

Although there is an abundance of information on gene expression during desiccation of seeds and zygotic embryos, there have been no published attempts to elucidate changes in gene expression during desiccation or the induction of desiccation tolerance of somatic embryos. Because the cell membranes are believed to be the primary site of desiccation injury, we isolated microsomal membrane proteins of alfalfa somatic embryos at desiccation-intolerant and -tolerant stages of development and compared their polypeptide profiles.[82] Although the level of two peptides at 53 and 32 kD, respectively, decreased with the onset of desiccation tolerance, there were no similarities between these peptides and microsomal membrane peptides associated with the onset of tolerance in seeds. In addition, the 53 and 32 kD peptides decreased as desiccation tolerance was established, whereas in other species peptides are synthesized or increase concurrent with the onset of desiccation tolerance.[55,81]

LEA Proteins

Dure and coworkers[57,58] have identified five classes of mRNAs that were regulated throughout seed formation, maturation, and desiccation. Of the five groups only one became abundant in late embryogenesis (after cell division had halted), but it also rapidly disappeared upon imbibition. The 14 members of this subset in cotton have since been well-characterized and have become known as the LEA or late-embryogenesis-abundant mRNAs coding for the LEA polypeptides.

The timing and structure of the LEA proteins are suggestive of a protective role during desiccation. They appear as desiccation begins and decline within 48 h of imbibition.[83–85] Hypothetically, the proteins interact with the membrane during dehydration in a type of water replacement, with the hydroxyl groups forming hydrogen bonds with the water-binding proteins or lipids within the membrane. As well, the amphiphilic helices may anchor the LEAs into the membrane matrix. Another postulation is that LEA proteins act as salt bridges between amino acids of charged proteins when the ionic strength of the cell increases during desiccation.[85] Currently, there are no studies examining the LEA proteins in somatic embryos, which is surprising since these peptides could play such a pivotal role in desiccation tolerance.

Summary

The induction of desiccation tolerance in somatic embryos is a critical prerequisite for drying somatic embryos as artificial seeds. Desiccation provides not only for both short- and long-term storage of these seed analogs, but it is presumed that it will facilitate the production of vigorous, high quality artificial seeds by emulating the maturation process of zygotic embryos. To date, desiccation tolerance has been induced to varying degrees in somatic embryos by exogenous application of ABA in the maturation medium or by exposure to sublethal stress that presumably stimulates ABA synthesis.

The onset of desiccation tolerance and subsequent desiccation causes extensive biochemical and physiological changes at the cellular level, including changes in carbohydrate and protein accumulation and in water binding characteristics. The accumulation of lipid soluble antioxidants and the reduction in the potential for activated oxygen production seem to be associated not only with the initial tolerance of water loss, but also to longevity in the dry state. The processes involved in zygotic embryo maturation are not sufficiently understood to allow these processes to be replicated in somatic embryos. This is currently a major limitation in the commercial production of artificial seeds and the application of this new technology.

Acknowledgments

The authors gratefully acknowledge the financial assistance of the Natural Sciences and Engineering Council of Canada (Strategic Grant), the University Research Incentive Fund of the province of Ontario, and Somatica Plant Technologies in the conduct of our research. Susan Van Acker is a recipient of the McConkey Scholarship for postgraduate studies.

References

1. **Gray, D. J. and Purohit, A.,** Somatic embryogenesis and development of synthetic seed technology, *Crit. Rev. Plant Sci.*, 10, 33, 1991.

2. **McKersie, B. D., Senaratna, T., Bowley, S. R., Brown, D. C. W., Krochko, J. E., and Bewley, J. D.,** Application of artificial seed technology in the production of hybrid alfalfa (*Medicago sativa* L.), *In Vitro Cell. Dev. Biol.*, 25, 1183, 1989.

3. **Dure, L. S., III,** Seed formation, *Annu. Rev. Plant Physiol.*, 26, 259, 1975.

4. **Walbot, V.,** Control mechanisms for plant embryogeny, in *Dormancy and Developmental Arrest*, Clutter, M. E., Ed., Academic Press, New York, 1978, 113.

5. **McKersie, B. D. and Bowley, S. R.,** Synthetic seeds of alfalfa, in *Application of Synthetic Seeds to Crop Improvement*, Redenbaugh, K., Ed., CRC Press, Boca Raton, FL, 1992, 231.

6. **Carman, J. G.,** Embryogenic cells in plant tissue cultures: occurrence and behaviour, *In Vitro Cell. Dev. Biol.,* 26, 746, 1990.

7. **Xu, N., Coulter, K. M., Krochko, J. E., and Bewley, J. D.,** Morphological stages and storage protein accumulation in developing alfalfa (*Medicago sativa* L.) seeds, *Seed Sci. Res.,* 1, 119, 1991.

8. **Ammirato, P. V.,** Organization events during somatic embryogenesis, in *Plant Tissue and Cell Culture,* Green C. E., Somers, D. A., Hackett, W. P., and Biesboer, D. D., Eds., Alan R. Liss, New York, 1987, 57.

9. **Stuart, D. A. and Nelsen, J.,** Isolation and characterization of alfalfa 7S and 11S seed storage protein, *J. Plant Physiol.,* 132, 129, 1988.

10. **Stuart, D. A., Nelsen, J., and Nichol, J. W.,** Expression of 7S and 11S alfalfa seed storage proteins in somatic embryos, *J. Plant Physiol.,* 132, 134, 1988.

11. **Krochko, J. E., Charbonneau, M. R., Coulter, K. M., Bowley, S. R., and Bewley, J. D.,** A comparison of seed storage protein in subspecies and cultivars of *Medicago sativa, Can. J. Bot.,* 68, 940, 1990.

12. **Krochko, J. E. and Bewley, J. D.,** Identification and characterization of the seed storage proteins from alfalfa (*Medicago sativa*), *J. Exp. Bot.,* 41, 505, 1990.

13. **Coulter, K. M. and Bewley, J. D.,** Characterization of a small sulphur-rich storage albumin in seeds of alfalfa (*Medicago sativa* L.), *J. Exp. Bot.,* 41, 1541, 1990.

14. **Krochko, J. E., Pramanik, S. K., and Bewley, J. D.,** Contrasting storage protein synthesis and messenger RNA accumulation during development of zygotic and somatic embryos of alfalfa (*Medicago sativa* L.), *Plant Physiol.,* 99, 46, 1992.

15. **Lai, F. M., Senaratna, T., and McKersie, B. D.,** Glutamine enhances storage protein synthesis in *Medicago sativa* L. somatic embryos, *Plant Sci.,* 87, 69, 1992.

16. **Hakman, I., Stabel, P., Engstrom, P., and Eriksson, T.,** Storage protein accumulation during zygotic and somatic embryo development in *Picea abies* (Norway Spruce), *Physiol. Plant,* 80, 441, 1990.

17. **McCleary, B. V. and Matheson, N. K.,** α-D-Galactosidase activity and galactomannan and galactosylsucrose oligosaccharide depletion in germinating legume seeds, *Phytochemistry,* 13, 1747, 1974.

18. **McCleary, B. V. and Matheson, N. K.,** Galactomannan utilization in germinating legume seeds, *Phytochemistry,* 15, 43, 1976.

19. **Singh, R., Kaur, P., Goyal, J., and Gupta, A. K.,** Interconversion and translocation of free sugars during galactomannan utilization in germinating guar (*Cyamopsis tetragonologa*) seed, *Plant Sci.,* 51, 21, 1987.

20. **Kohno, A., Shinke, R., and Nanmori, T.,** Features of the β-amylase isoform system in dry and germinating seeds of alfalfa (*Medicago sativa* L.), *Biochim. Biophys. Acta,* 1035, 325, 1990.

21. **Fujii, J. A. A., Slade, D., Olsen, R., Ruzin, S. E., and Redenbaugh, K.,** Alfalfa somatic embryo maturation and conversion to plants, *Plant Sci.,* 72, 93, 1991.

22. **Lai, F. M. and McKersie, B. D.,** unpublished.

23. **Fountain, D. W. and Outred, H. A.,** Seed development in *Phaseolus vulgaris* L. cv Seminole. II. Precocious germination in late maturation, *Plant Physiol.,* 93, 1089, 1990.

24. **Hong, T. D. and Ellis, R. H.,** A comparison of maturation drying, germination, and desiccation tolerance between developing seeds of *Acer pseudoplatanus* L. and *Acer platanoides* L., *New Phytol.,* 116, 589, 1990.

25. **Kermode, A. R. and Bewley, J. D.,** The role of maturation drying in the transition from seed development to germination, *J. Exp. Bot.,* 36, 1906, 1985.

26. **Galau, G. A., Jakobsen, K. S., and Hughes, D. W.,** The controls of late dicot embryogenesis and early germination, *Physiol. Plant.,* 81, 280, 1991.

27. **Leopold, A. C.,** Coping with desiccation, in *Stress Responses in Plants: Adaptation and Acclimation Mechanisms,* Alscher, R. G. and Cumming, J. R., Eds., Alan R. Liss, 1990, 37.

28. **Welbaum, G. E. and Bradford, K. J.,** Water relations of seed development and germination in muskmelon (*Cucumis melo* L.). II. Development of germinability, vigour, and desiccation tolerance, *J. Exp. Bot.,* 40, 1355, 1989.

29. **Karssen, C. M., Brinkhorst-vanderSwan, D. L. C., Breekland, A. E., and Koorneef, M.,** Induction of dormancy during seed development by endogenous abscisic acid: studies on abscisic acid deficient genotypes of *Arabidopsis thaliana* (L.) Heynh, *Planta,* 157, 158, 1983.

30. **Senaratna, T., McKersie, B. D., and Bowley, S. R.,** Desiccation tolerance of alfalfa (*Medicago sativa* L.) somatic embryos. Influence of abscisic acid, stress pretreatments and drying rates, *Plant Sci.,* 65, 253, 1989.

31. **Senaratna, T., McKersie, B. D., and Bowley, S. R.,** Artificial seeds of alfalfa (*Medicago sativa* L.). Induction of desiccation tolerance in somatic embryos, *In Vitro Cell. Dev. Biol.,* 26, 85, 1990.

32. **Anandarajah, K. and McKersie, B. D.,** Manipulating the desiccation tolerance and vigor of dry somatic embryos of *Medicago sativa* L. with sucrose, heat shock and abscisic acid, *Plant Cell Rep.,* 9, 451, 1990.

33. **Marsolais, A. A., Wilson, D. P. M., Tsujita, M. J., and Senaratna, T.,** Somatic embryogenesis and artificial seed production in Zonal (*Pelargonium x hortorum*) and Regal (*Pelargonium x domesticum*) geranium, *Can. J. Bot.,* 69, 1188, 1991.

34. **Black, M.,** Involvement of ABA in the physiology of developing and mature seeds, in *Abscisic Acid Physiology and Biochemistry,* Davies, W. J. and Jones, H. J., Eds., Bios Scientific Publishers, Oxford, 1991, 99.

35. **Kitto, S. L. and Janick, J.,** Hardening treatments increase survival of synthetically-coated asexual embryos of carrot, *J. Am. Soc. Hortic. Sci.,* 110, 283, 1985.

36. **Roberts, D. R., Sutton, B. C. S., and Flinn, B. S.,** Synchronous and high frequency germination of interior spruce somatic embryos following partial drying at high relative humidity, *Can. J. Bot.,* 68, 1086, 1990.

37. **Roberts, D. R., Lazaroff, W. R., and Webster, F. B.,** Interaction between maturation and high relative humidity treatments and their effects on germination of Sitka Spruce somatic embryos, *J. Plant Physiol.,* 138, 1, 1991.

38. **Attree, S. M., Moore, D., Sawhney, V. K., and Fowke, L. C.,** Enhanced maturation and desiccation tolerance of White Spruce [*Picea glauca* (Moench) Voss] somatic embryos: effects of a non-plasmolysing water stress and abscisic acid, *Ann. Bot.,* 68, 519, 1991.

39. **Attree, S. M., Pomeroy, M. K., and Fowke, L. C.,** Manipulation of conditions for the culture of somatic embryos of White Spruce for improved triacylglycerol bio-synthesis and desiccation tolerance, *Planta,* 187, 395, 1992.

40. **Kim, Y. H. and Janick, J.,** Synthetic seed technology: improving desiccation tolerance of somatic embryos of celery, *Acta Hortic.,* 280, 23, 1990.

41. **Gray, D. J.,** Quiescence in monocotyledonous and dicotyledonous somatic embryos induced by dehydration, *HortScience,* 22, 810, 1987.

42. **Hetherington, A. M. and Quatrano, R. S.,** Mechanisms of action of abscisic acid at the cellular level, *New Phytol.,* 119, 9, 1991.

43. **Perata, P., Picciarelli, P., and Alpi, A.,** Pattern of variations in abscisic acid content in suspensors, embryos, and integuments of developing *Phaseolus coccineus* seeds, *Plant Physiol.,* 94, 1776, 1990.

44. **Xu, N., Coulter, K. M., and Bewley, J. D.,** Abscisic acid and osmoticum prevent germination of developing alfalfa embryos, but only osmoticum maintains the synthesis of developmental proteins, *Planta,* 182, 382, 1990.

45. **Oishi, M. Y. and Bewley, J. D.,** Distinction between the responses of developing maize kernels to fluridone and desiccation in relation to germinability, α-amylase activity, and abscisic acid content, *Plant Physiol.,* 94, 592, 1990.

46. **Finkelstein, R. R., Tenbarge, K. M., Shumway, J. E., and Crouch, M. L.,** Role of ABA in maturation of rapeseed embryos, *Plant Physiol.,* 78, 630, 1985.

47. **Arnold, R. L. B., Fenner, M., and Edwards, P. J.,** Changes in germinability, ABA content and ABA embryonic sensitivity in developing seeds of *Sorghum bicolor* (L.) Moench. induced by water stress during grain filling, *New Phytol.,* 118, 339, 1991.

48. **Welbaum, G. E., Tissaoui, T., and Bradford, K. J.,** Water relations of seed development and germination in muskmelon (*Cucumis melo* L.). III. Sensitivity of germination to water potential and abscisic acid during development, *Plant Physiol.,* 92, 1029, 1990.

49. **Ried, J. L. and Walker-Simmons, M. K.,** Synthesis of abscisic acid- responsive, heat-stable proteins in embryonic axes of dormant wheat grain, *Plant Physiol.,* 93, 662, 1990.

50. **Kim, Y. H. and Janick, J.,** Abscisic acid and proline improve desiccation tolerance and increase fatty acid content of celery somatic embryos, *Plant Cell Tissue Organ Cult.,* 24, 83, 1991.

51. **Anandarajah, K., Kott, L., Beversdorf, W. D., and McKersie, B. D.,** Induction of desiccation tolerance in microspore-derived embryos of *Brassica napus* L. by thermal stress, *Plant Sci.,* 77, 119, 1991.

52. **Skriver, K. and Mundy, J.,** Gene expression in response to abscisic acid and osmotic stress, *Plant Cell,* 2, 503, 1990.

53. **McKersie, B. D., Senaratna, T., and Bowley, S. R.,** Drying somatic embryos for use as artificial seed, *Proc. Plant Growth Reg. Soc.,* 17, 199, 1990.

54. **Parrott, W. A., Dryden, G., Vogt, S., Hinderbrand, D. F., Collins, G. B., and Williams, E. G.,** Optimization of somatic embryogenesis and embryo germination in soybean, *In Vitro Cell. Dev. Biol.,* 24, 817, 1988.

55. **Lane, B. G.,** Cellular desiccation and hydration: developmentally regulated proteins, and the maturation and germination of seed embryos, *FASEB J.,* 5, 2893, 1991.

56. **Bartels, D., Schneider, K., Terstappen, G., Piatkowski, D., and Salamini, F.,** Molecular cloning of abscisic acid-modulated genes which are induced during desiccation of the resurrection plant *Craterostigma plantagineum, Planta,* 181, 27, 1990.

57. **Galau, G. A., Hughes, D. W., and Dure, L., III,** Abscisic acid induction of cloned cotton late embryogenesis-abundant (Lea) mRNAs, *Plant Mol. Biol.,* 7, 155, 1986.

58. **Dure, L., III, Crouch, M., Harada, J., Ho, T. D., Mundy, J., Quatrano, R., Thomas, T., and Sung, Z. R.,** Common amino acid sequence domains among the LEA proteins of higher plants, *Plant Mol. Biol.,* 12, 475, 1989.

59. **Bewley, J. D. and Black, M.,** *Physiology and Biochemistry of Seeds in Relation to Germination. I. Development, Germination and Growth,* Springer-Verlag, New York, 1978, 306.

60. **D'Arcy, R. L. and Watt, I. C.,** Analysis of sorption isotherms of non-homogeneous sorbents, *Trans. Faraday Soc.,* 66, 1236, 1970.

61. **Vertucci, C. W. and Leopold, A. C.,** Bound water in soybean seed and its relation to respiration and imbibitional damage, *Plant Physiol.,* 75, 114, 1984.

62. **Vertucci, C. W. and Leopold, A. C.,** Water binding in legume seeds, *Plant Physiol.,* 85, 224, 1987.

63. **Vertucci, C. W. and Leopold, A. C.,** The relationship between water binding and desiccation tolerance in tissues, *Plant Physiol.,* 85, 232, 1987.

64. **Senaratna, T., McKersie, B. D., and Stinson, R. H.,** Association between membrane phase properties and dehydration injury in soybean axes, *Plant Physiol.,* 76, 759, 1984.

65. **Reid, J. S. G.,** Cell wall storage carbohydrates in seeds — biochemistry of the seed "gums" and "hemicelluloses", *Adv. Bot. Res.,* 11, 125, 1985.

66. **Crowe, J. H.,** Anhydrobiosis: an unsolved problem, *Am. Nat.,* 105, 563, 1971.

67. **Crowe, L. M. and Crowe, J. H.,** Effects of water and carbohydrates on membrane fluidity, in *Physiological Regulation of Membrane Fluidity,* Aloia, R. C., Curtain, C. C., and Gordon, L. M., Eds., Alan R. Liss, New York, 1988, 75.

68. **Clegg, J. S.,** The physical properties and metabolic status of *Artemia* cysts at low water contents: the "water replacement hypothesis", in *Membranes, Metabolism, and Dry Organisms,* Leopold, A. C., Ed., Comstock Publishing, Ithaca, 1986, 169.

69. **Hoekstra, F. A. and van Roekel, T.,** Desiccation tolerance of *Papaver dubium* L. pollen during its development in the anther. Possible role of phospholipid composition and sucrose content, *Plant Physiol.,* 88, 626, 1988.

70. **Simon, E. W.,** Phospholipids and plant membrane permeability, *New Phytol.,* 73, 337, 1974.

71. **Amuti, K. S. and Pollard, C. J.,** Soluble carbohydrates of dry and developing seeds, *Phytochemistry,* 16, 529, 1977.

72. **Bianchi, G., Murelli, C., Bochicchio, A., and Vazzana, C.,** Changes of low-molecular weight substances in *Boea hygroscopia* in response to desiccation and rehydration, *Phytochemistry,* 30, 461, 1991.

73. **Koster, K. L.,** Glass formation and desiccation tolerance in seeds, *Plant Physiol.,* 96, 302, 1991.

74. **Senaratna, T. and McKersie, B. D.,** Dehydration injury in germinating soybean (*Glycine max* L. Merr) seeds, *Plant Physiol.,* 72, 620, 1983.

75. **Senaratna, T., McKersie, B. D., and Stinson, R. H.,** Simulation of dehydration injury to membranes from soybean axes by free radicals, *Plant Physiol.,* 77, 472, 1985.

76. **Senaratna, T., McKersie, B. D., and Stinson, R. H.,** Antioxidant levels in germinating soybean seed axes in relation to free radical and dehydration tolerance, *Plant Physiol.,* 78, 168, 1985.

77. **Senaratna, T., Gusse, J. F., and McKersie, B. D.,** Age-induced changes in the cellular membranes of inbibed soybean seeds, *Physiol. Plant.,* 73, 85, 1988.

78. **Elstner, E. F.,** Oxygen activation and oxygen toxicity, *Annu. Rev. Plant Physiol.,* 33, 73, 1982.

79. **Mundy, J. and Chua, N. H.,** Abscisic acid and water-stress induce the expression of a novel rice gene, *EMBO J.,* 7, 2279, 1988.

80. **Gomez, J., Sanchez-Martinez, D., Stiefel, V., Rigau, J., Puidomenech, P., and Pages, M.,** A gene induced by the plant hormone abscisic acid in response to water stress encodes a glycine-rich protein, *Nature,* 334, 262, 1988.

81. **Blackman, S. A., Wettlaufer, S. H., Obendorf, R. L., and Leopold, A. C.,** Maturation proteins associated with desiccation tolerance in soybean, *Plant Physiol.,* 96, 868, 1991.

82. **Van Acker, S. D. N. and McKersie, B. D.,** unpublished.

83. **Galau, G. A., Bijaisoradat, N., and Hughes, D. W.,** Accumulation kinetics of cotton late embryogenesis-abundant mRNAs and storage protein mRNAs: coordinate regulation during embryogenesis and the role of absicic acid, *Dev. Biol.,* 123, 198, 1987.

84. **Dure, L., III, Greenway, S. C., and Galau, G. G.,** Developmental biochemistry of cottonseed embryogenesis and germination: changing messenger ribonucleic acid populations as shown by *in vitro* and *in vivo* protein synthesis, *Biochemistry,* 20, 4162, 1981.

85. **Baker, J., Steele, C., and Dure, L., III,** Sequence and characterization of 6 Lea proteins and their genes from cotton, *Plant Mol. Biol.,* 11, 277, 1988.

Reactor Design for Plant Cell Suspension Culture

Gurmeet Singh[1] and Wayne R. Curtis[1,2]

[1]*Department of Chemical Engineering, The Pennsylvania State University, University Park, Pennsylvania*

[2]*Biotechnology Institute, The Pennsylvania State University, University Park, Pennsylvania*

Introduction

Plants have developed tremendous capabilities to synthesize chemicals to enhance their competitiveness within the environment. Since the beginning of recorded time mankind has exploited these capabilities for their own benefit as sources of flavors, dyes, pesticides, and pharmaceuticals. However, harvesting these chemicals from either natural or cultivated sources is subject to many problems that range from ecological to political. In the future, we will undoubtedly rely on chemicals that are produced from plants which grow in remote and/or threatened ecosystems — such as the rain forests — which make agronomic cultivation impossible and destructive harvest unacceptable. The ability to grow plant cells in axenic tissue culture provides an attractive alternative to harvesting chemicals from whole plants. Within the environment of a reactor, the biochemical capabilities of plants can be exploited independent of political unrest or environmental regulation.

From a reaction engineering perspective, a suspended plant cell can be viewed as a biological catalyst — analogous to microorganisms for antibiotic formation or human hybridomas for monoclonal antibody production. Despite

0-8493-8262-9/94/$0.00+$.50

this superficial similarity, plant cells have numerous unique characteristics that must be considered to achieve large-scale commercial culture. Many of the problems are similar to those initially faced in mammalian cell culture, such as shear sensitivity and erratic growth behavior. These problems resulted in a tremendous variety of reactor designs for mammalian cell culture including hollow fibers, macroporous beads and supports, and immobilized and fluidized beds. These novel designs must compete with modifications in the conventional stirred-tank fermenter which has proven its effectiveness in the microbial 'fermentation' industry. In the case of small volume production, such as antibody formation for diagnostics, there is room for novelty since the ultimate scale of production is comparatively small and there is little or no need for scale-up. On the other hand, for products which are therapeutic or commodity chemicals, the ultimate scale of production will be tens of thousands of liters. Under these circumstances, scalability is of primary concern, and the ability to implement production in existing technology with minor modifications is very attractive. Most of the chemicals of interest from plant tissue culture fall into this latter category where large scale production is a necessary component of economic feasibility. Along these lines, this chapter takes the perspective that the 'industry standard', mechanically agitated deep-tank fermenter is the appropriate starting point from which modifications and alternative designs should be justified for the cultivation of plant cell suspensions.

Progress on Growth of Plant Cells in Agitated Reactors

A wide array of reactors has been used to grow plant cell suspension cultures on a large scale (Table 1).[1–9] Most initial work has been conducted in stirred-tank and airlift reactors. Stirred tanks used for microbial cultures are designed to operate at high power input per unit volume to provide adequate oxygen transfer for rapid growth. When plant cells were initially grown in these systems they showed poor growth, which was interpreted as high shear sensitivity for plants.[10] This led to a shift in focus to airlift reactors. Wagner and Vogelmann[7] did a comparative study of airlift and stirred-tank reactors and concluded that airlift reactors were better for plant cell suspensions due to their lower shear hydrodynamics. Various attempts were made at culturing plant cells in larger and larger airlift reactors. Smart and Fowler[11] operated a 10-l airlift bioreactor to grow *Catharanthus roseus* cell suspensions and later scaled it up to 85 l. Reinhard and Kreis[12] cultured *Digitalis lanata* in airlift reactors up to 300 l in volume.

When plant cell suspensions were cultured in airlift bioreactors, difficulties were encountered in providing adequate mixing for cultures at high cell concentrations.[3,13] Other inhibitory effects, manifested as reduced growth rates, were also observed at the high sparge rates required for pneumatic agitation.[14] The problems encountered with airlift reactors, combined with increased scale-up

Table 1 *Various types of bioreactors used for culturing plant cell suspensions*

Species	Reactor type	Working volume (l)	Special features	Tissue density (g DW/l)	Culture time (d)	Ref.
Catharanthus roseus	STR	8.5		13	7	1
	Airlift	30		17.2	9	2
	Airlift	30	Fed batch	23.4	16	2
	RDF	5		20	19	3
Cudriana tricuspidata	STR	5	Paddle impeller	30	25	3
Dioscorea deltoidea	STR	10	Pitched turbine	8.0	10	4
Echinacea purpurea	STR	75,000	INTERMIG impeller	200*	—	5
Hyoscyamus muticus	STR	10	ABEC impeller™ *	9.0	14	6
Morinda citrifolia	Airlift	10		15	7	7
Nicotiana tabacum	STR	15,500		15	3.3	8
Thalictrum rugosum	STR	5	Cell lift impeller, perfusion reactor	31	36	9

STR = stirred-tank reactor; RDF = rotating drum fermenter.
* ABEC low shear impeller design, Elephant Ear Impeller, is a trademark of A.B.E. Company, Allentown, PA. ** Tissue density expressed in g FW/l.

success in conventional stirred-tank fermenters,[8] shifted the emphasis in scale-up back to mechanically agitated reactors. The major modifications of conventional fermenters which have been employed for plant cell suspension cultures have been agitator and sparger design. The goal of these changes is to reduce shear while at the same time provide sufficient oxygen transfer and mixing. A cell lift impeller (New Brunswick, designed for mammalian cell culture) was used by Treat et al.[15] to achieve viability of 90% and a cell aggregate size <1 mm in vessels up to 2.5 l in volume. By comparison, flat blade turbines displayed a reduced viability and larger aggregate size. Hooker et al.[16] used 5-l fermenters equipped with four-bladed impellers of large widths to reduce cell shear damage. Leckie et al.[1] utilized a pitch-blade turbine and showed that both the growth and the yield increased relative to the conventional Rushton turbine impeller in 12-l stirred-tank bioreactors. Jolicoer et al.[17] used a low shear, double helical ribbon impeller to grow high density *Catharanthus roseus* cell suspensions in an 11-l stirred-tank bioreactor.

The tendency toward lower shear conditions results in a reduced dispersion of gas within the reactor. More specifically, to minimize the likelihood of disrupting the cells, the operating conditions are no longer conducive to bubble break-up. This reduction in impeller dispersion of gas has been countered through the use of fine-bubble sparging to enhance interfacial area for mass transfer. Leckie et al.[18] used sintered stainless-steel spargers to achieve fine bubbles compared to the cross spargers. Kim et al.[9] used a hybrid bioreactor, which was essentially a stirred-tank reactor with a cell-lift impeller and a sintered metal sparger, to achieve cell concentrations of 31 g dry weight (DW) per liter. Ritterhaus et al.[5] operated a 75,000-l stirred-tank reactor fitted with INTERMIG stirrer, in which the stirring cell covers the full height and is not divided into individual stirring cells as in the case with blade stirrers. They were able to grow *Echinacea purpurea* to a cell concentration of over 200 g/l (fresh weight basis or FW), thus putting beyond doubt the ability to culture plant cell suspensions on a large scale in a stirred tank.

Some attempts at scale-up have considered alternate reactor types such as the immobilized cell reactor,[19] cell membrane reactor,[20] magnetically stabilized fluidized bed,[21] and rotating drum fermenter.[3] However, for the purpose of this chapter we will limit our discussion to the stirred-tank reactor since it appears to hold the maximum potential for successful scale-up.

Bioreactor Design

Bioreactors for plant cell suspension culture have three phases: the liquid media liquid phase, the solid plant cell aggregates, and the gas phase, which is usually sparged air. The reactor design must supply sufficient power input to provide adequate levels of liquid mixing and oxygen transfer from the gas to the liquid phase. In addition, the level of mixing must maintain plant cell aggregates in suspension. Mechanical agitation, however, poses both positive and negative effects. Since the detrimental effects of mixing can be observed at relatively mild operating conditions, the desire to increase productivity will result in bioreactor operation at conditions very close to the point at which cells are stressed. Under these conditions, scale-up can be a particularly difficult challenge due to shifts in the influence of factors governing mixing shear and oxygen transfer in larger stirred tank reactors.

Positive Effects of Power Input to the Bioreactor

Mixing

Mixing is required to maintain homogeneity of liquid and solid phases. The intensity of mixing in a stirred tank is usually characterized by the mixing time which is correlated by equations of the following functional form:

$$t_m = \frac{K}{N} \frac{\left(\frac{D_T}{D}\right)^{0.3}}{N_P^{\ 0.33}} \tag{1}$$

At relatively low cell concentrations (<200 g FW/l), the mixing time for a stirred-tank reactor is of the order of seconds. By comparison, the medium surrounding the cells contains sufficient carbon source (sugar) and inorganic salts to sustain growth for several weeks. As with microbial cultures, oxygen is present at concentrations which are low in comparison to the rate of consumption. As a result, the effects of mixing on design are usually discussed in the context of the effects of mixing on oxygen mass transfer as opposed to providing homogeneity within the liquid phase. Due to the relatively low rates of growth of plant tissue in culture, oxygen availability becomes a concern only at high cell concentrations. The low biological oxygen demand of plant cell cultures permits growth of suspensions to solids concentrations that can occupy as much as 40% of the culture volume. This means that the balancing of oxygen transfer will take place at much higher cell concentrations, where mixing plays an important role in overall rates of oxygen mass transfer.

As the concentration of cells in the culture medium increases, the viscosity increases due to the displacement of the medium by the biotic phase. Many studies have concluded that plant cell cultures are highly viscous; however, we recently reported in a rather comprehensive study that the viscosity of plant cell suspensions is comparable to other microbial suspensions at the same biotic volume.[22] For most plant cell suspensions, the apparent viscosity remains relatively low (<15 cP) up to a cell concentration of 200 g FW/l. At higher concentrations, the viscosity increases rapidly until it reaches a solid mass that can be proliferated in small scale as an undifferentiated tumorous tissue called callus. In practice, the cells must be kept at a concentration low enough to retain fluidity; however, it is clear that at sufficiently high cell concentrations, the characteristic time of mixing will become comparable to the characteristic times of mass transfer (on the order of minutes). Moreover, stagnant zones can result in the reactor at high cell densities, disrupting the homogeneity of the reactor. This can result in nutrient and temperature gradients and poor control. Mixing may become the governing criterion of design under these circumstances.

For aggregated suspensions, the intensity of required mixing may be determined by the rate of sedimentation of the plant cell aggregates. Figure 1 shows the settling velocities of cell aggregates of various plant species. The settling velocities are found to range from 0.1 to 10 cm/s and Reynolds number (N_{Re}) ranging from <1 to around 1000. For plant cell aggregates with $N_{Re} < 2$ (aggregate sizes of the order of 0.1 mm or less), sedimentation will be viscosity-dependent. This flow regime is referred to as hindered settling. The role of hindered settling will become particularly important in high density culture

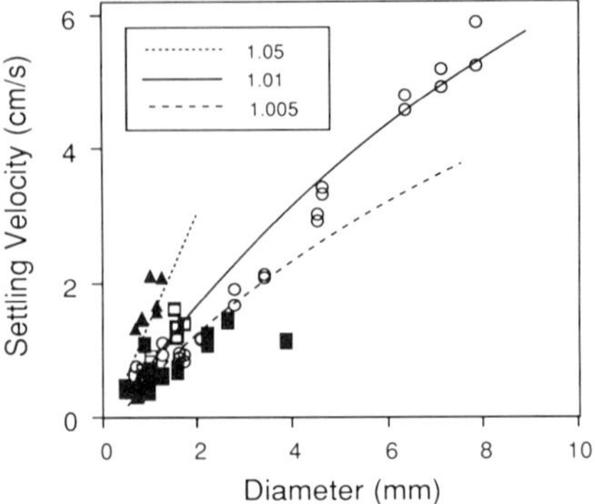

Figure 1. *Settling velocities of cell aggregates of various plant species:* ○ = *neem,* ■ = *potato,* □ = *carrot,* ▲ = *oak embryo culture. The contours represent theoretically calculated settling velocities for constant density spherical particles (as indicated in the figure inset).*

where the increased culture viscosity decreases the circulation rates within the vessel. This reduced circulation reduces the ability to suspend cell aggregates; however, this is offset by the reduced rates of sedimentation resulting from increased particle interaction at high cell concentrations.

As the rate of sedimentation increases for larger cell aggregates, the dependence of sedimentation rate on viscosity decreases. The transition region from viscosity-dependent to viscosity-independent sedimentation falls in a N_{Re} range of 2 to 500. Above a N_{Re} of 500 (aggregate sizes 1 to 10 mm and larger), the sedimentation will become viscosity-independent. Since a plant cell culture contains a wide range of plant cell aggregate sizes, the smaller (usually predominant) aggregates will increase the viscosity and decrease culture circulation rates. Since the sedimentation rates of larger aggregates are not affected by this increase in viscosity, they will have a greater tendency to settle to the bottom of the reactor. Figure 2 shows the bottom of a 5-l mechanically agitated vessel after 18 d of culture of an aggregated plant cell line. The operational conditions, including an impeller speed of 150 rpm, were more than adequate for the oxygen requirements of the culture; however, as shown in the photograph, the bottom of the reactor is covered with large aggregates which had settled out and stagnated at the bottom of the vessel as the run progressed. The stagnant zones between the large aggregates then accumulated the smaller aggregates, resulting in a solid layer of plant aggregates about an inch thick.

Aggregate accumulation on the bottom of the vessel can be avoided if the impeller speed is sufficiently high. Numerous correlations are available for calculating the minimum critical impeller speed, N_{Cr}, required for off-bottom suspension of particles. Although these correlations are usually regressed based on data at relatively low particle loadings, they provide a reasonable basis for estimating the conditions required to suspend plant cell aggregates.

Figure 2. *A 5-l mechanically agitated vessel showing large aggregates of cells settled at the bottom. The culture time was 18 d and the impeller speed during operation was maintained at 150 rpm.*

N_{Cr} is dependent upon fluid properties such as kinematic viscosity (ν) and density (ρ), reactor dimensions including impeller diameter (D), particle weight percent (B), particle diameter (D_p), and density (ρ_p). When these correlations are applied to a small scale (10 l) vessel for properties of a 'typical' plant cell culture, N_{Cr} varies in the range of 50 to 250 rpm, depending on the correlation selected. This range of impeller rpm spans the normal operating range for plant cell culture. One representative correlation for critical impeller speed is given below:[23]

$$N_{Cr} = K(\nu)^{0.1}(D_P)^{0.2}\left(g\frac{\rho_P - \rho}{\rho}\right)^{0.45}(B)^{0.13}(D)^{-0.85} \qquad (2)$$

where the proportionality constant K depends upon the units chosen for the parameters. This equation was used to calculate the aggregate suspension conditions required during scale-up of a 'typical' plant cell suspension at 200 g FW/l and an aggregate specific gravity of 1.05 (Figure 3). The minimum impeller speed required to suspend an aggregate of diameter D_p decreases at larger reactor volumes, but the impeller tip speed which corresponds to the maximum shear rate increases (Figure 3A). This enhancement in suspension

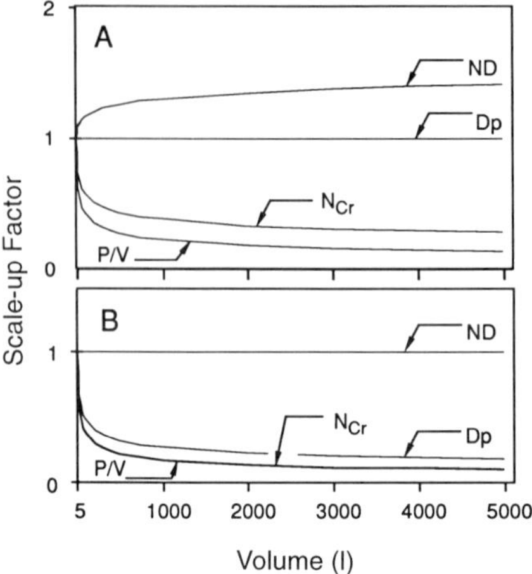

Figure 3. *Effect of scale-up based on critical speed for off-bottom aggregate suspension (Equation 2). Panel A shows effects of impeller tip speed (ND), impeller speed (N), and power drawn per unit volume (P/V) at constant aggregate diameter, D_P. Panel B shows the implications of 'typical' scale-up based on constant impeller tip speed, ND. The analysis was based on cell suspensions at concentrations of 200 g FW/l grown in a 5-l vessel, $L/D_T = 1.5$, $D/D_T = 1/3$, D = 6.2 cm (axial flow impeller), tip speed = 0.5 m/s (impeller speed of 2.61 rps).*

results from increased flow achieved by larger impellers. It should be noted, however, that upon scale-up, reactors are usually taller with multiple impellers to minimize torque on drive motor. As a result, in conventional reactor geometries, an enhancement in suspension capability will not likely be realized. As shown in Figure 3B, if scale-up is based on constant tip speed, the maximum aggregate size that can be suspended decreases. As a result, the scale-up of an aggregated cell line may be based on an interaction of suspension and shear as opposed to oxygen transfer and shear, which is the more familiar basis for scale-up. In the context of the effects of aggregate size on suspension, it should be noted that, in general, stress results in an increase in aggregate size. Extremely large aggregates have been noted upon scale-up of tobacco cultures — most likely due to the stresses associated with growth in a mechanically agitated system.[24] There is also speculation that increased aggregation may be beneficial for secondary metabolite formation. These observations should be kept within the context of the ultimate goal of large scale production.

Fortunately, many suspension cultures are not highly aggregated ($D_p \ll 1$ mm), particularly those which have been maintained by serial subculture for long periods of time. For these cultures, the relatively low sedimentation velocity combined with the viscosity-dependent 'hindered settling' enhances the ability to keep the culture well-mixed.

Interfacial Mass Transfer

Agitation is also a means of enhancing the oxygen supply to a culture in a conventional bioreactor. Since oxygen consumption is a fundamental culture requirement, any bioreactor design must consider the biological oxygen demand of plant cells. Based on stoichiometry of carbohydrate metabolism, an expression for the theoretical biological oxygen demand can easily be derived:

$$\mathrm{BOD} = 3.47\mu X \frac{\left(1 - \frac{\omega_t Y}{\omega_s}\right)}{\left(\frac{Y}{\omega_s}\right)} \tag{3}$$

where BOD is the biological oxygen demand (mmoles O_2/l/h), μ is the specific growth rate (day^{-1}), X is the cell concentration (g DW/l), Y is the cell yield (g DW of cells per gram of sugar), ω_t is the fraction of carbon in DW of tissue, and ω_s is the fraction of carbon in the sugar source.

In the case of plant cells, the maximum specific growth rates may vary from 0.1 to 0.5 day^{-1}. Based on a 'typical' carbon yield of 0.45, and 50% DW carbon, a surface can be generated that describes predicted BOD as a function of cell density and specific growth rate (Figure 4). An analysis of the oxygen uptake rate for *Hyoscyamus muticus* cell suspension culture was done by growing them in a 10-l New Brunswick Scientific Micros™* bioreactor. The oxygen uptake rates were calculated by the dynamic method, which involves stopping gas flow and measuring dissolved oxygen (DO) depletion.[25] The inlet and outlet gas streams were also analyzed with a quadrapole mass spectrometer. The oxygen uptake rate was then calculated based on the observed oxygen differential as well as the carbon dioxide differential. These experimental measurements of BOD are compared along with theoretical BOD based on Equation 3 calculated from cell density sampling. The maximum experimental oxygen uptake rate corresponds to the maximum theoretical BOD; however, the theoretical biological oxygen demand appears to be much greater than the experimental oxygen consumption rate (Figure 5). Review of the literature (Table 2) also reveals significant differences between experimentally measured biological oxygen demands of various plant cell cultures and the theoretical demand obtained from stoichiometry. This difference could be attributed to the large variations in biomass composition, which limit the applicability of overall stoichiometric coefficients.[28] Clearly, better estimates of BOD are needed due to the importance of these parameters to reactor design; nonetheless, these preliminary calculations and experimental measurements still provide an important starting point which can be refined as needed for specific

* Registered trademark of New Brunswick Scientific Company, Inc., Edison, NJ.

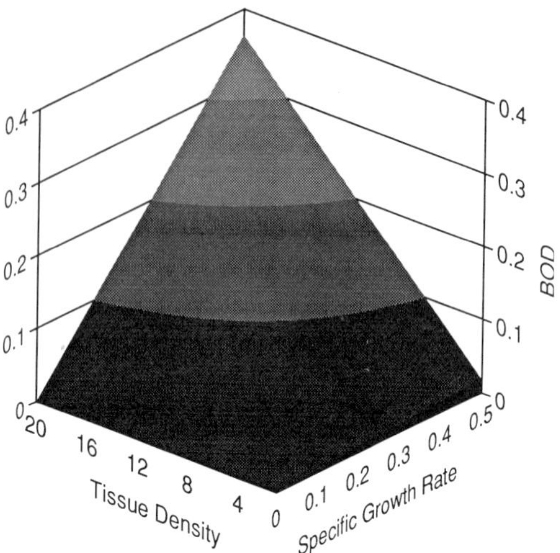

Figure 4. *Surface representing the theoretical biological oxygen demand (BOD) of a plant cell culture as a function of tissue density and specific growth rate (Equation 3). The BOD is represented in mmol/O_2/l, tissue density in gram DW/l and specific growth rate in (day)$^{-1}$.*

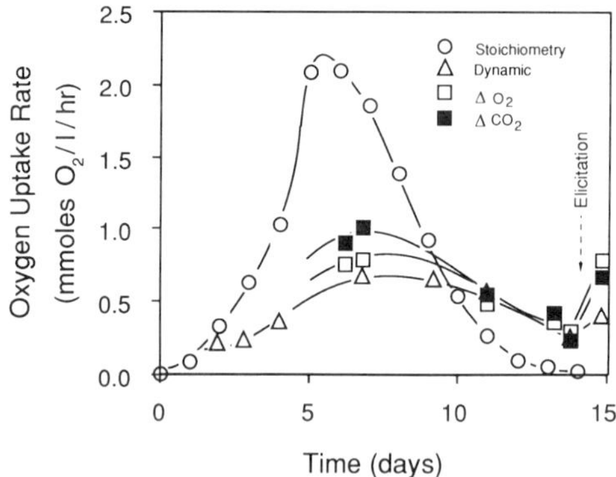

Figure 5. *Comparison of theoretical and experimentally obtained biological oxygen demand of* Hyoscyamus muticus *cell suspension cultures grown in a 10-l New Brunswick Micros™ fermenter. The theoretical estimate was obtained from Equation 3. $Y = 0.45$, $\omega_t = 0.5$, $\omega_s = 0.421$ for sucrose as the carbon source. Experimental estimates were obtained by dynamic degassing and gassing method of Pirt (Reference 25) as well as by mass balances for O_2 and CO_2 based on gas compositions of the inlet and outlet gas streams obtained from a quadrapole mass spectrometer.*

Table 2 *Comparison of experimentally obtained oxygen uptake rates and theoretical biological oxygen demands of a few plant cell suspension cultures*

Species	μ	X	OUR_{max}	BOD_{theo}	Y	Reactor type	Ref.
Catharanthus roseus	0.25	4	1.87	2.24	0.37	STR	4
Dioscorea deltoidea	0.166	6	1.87	2.44	0.34	STR	4
Hyoscyamus muticus	0.23	6	0.8	1.96	0.45	STR	6
Eschscholtzia californica	0.096	9	4.0	1.51	0.42	Shake flask	26
	0.06	15	2.83	1.07	0.5	Bubble column	
Fragario ananassa	0.13	13.3	1.33	2.8	0.44	Shake flask	27

Note: μ = Specific growth rate, day^{-1}; X = cell concentration based on dry weight, g/l; OUR = oxygen uptake rate, mmoles of O_2 per l/h; BOD_{theo} = theoretical biological oxygen demand, mmoles of O_2 per l/h; STR = stirred-tank reactor.

systems. As a minimum criterion for design, the overall mass transfer coefficient (k_la) of the system must be large enough to meet the culture BOD. The k_la for a bioreactor is defined in terms of the oxygen transfer rate (OTR):

$$OTR = k_la(DO^* - DO) \tag{4}$$

where DO is the operational dissolved oxygen level and DO* is the equilibrium dissolved oxygen concentration (which is about 0.25 mmol O_2 per l in equilibrium with air at 25°C).

If the rate of oxygen transfer is less than the oxygen demand of the culture, the liquid concentration of oxygen, DO, will be zero. However, the rate of respiration of microbial cultures is known to fall sharply below a liquid oxygen concentration of 0.003 to 0.05 mmol/l.[29] This DO level is known as the critical oxygen concentration. If the value of DO is greater than the critical oxygen value, the rate of oxygen utilization is limited by factors other than the oxygen transfer rate. Information on critical oxygen concentration for plant cell cultures are limited and contradictory. Tate and Payne[30] estimated the critical oxygen concentration of *Catharanthus roseus* as 0.12 mmol O_2 per l, which corresponds to 48% of saturation, whereas the estimate of Parrelieux and Vinas[31] was 0.027 mmol O_2 per l, which corresponds to about 12% of saturation. By monitoring the levels of NAD(P)H using excitation fluorometry, Asali et al.[32] observed that accumulation of NAD(P)H, indicating a metabolic starvation of oxygen, was not noticed until the DO level reached about 11% which is consistent with the measurement of Parrelieux and Vinas. A minimum k_la for culture growth can be calculated by setting the OTR equal to the BOD of the culture, and DO level equal to DO_{Cr}.

$$(k_l a)_{min} = \frac{BOD}{(DO^* - DO_{Cr})} \tag{5}$$

If the critical oxygen concentration is high, then significantly higher mass transfer coefficients will be required due to the reduced oxygen transfer driving force. On the other hand, if DO_{Cr} is small then a substantially lower $k_l a$ will be sufficient. Based on experimentally measured BOD for *Catharanthus roseus,* and the preceding development, it can be deduced that the minimum $k_l a$ required for cultivation of 'typical' plant cells will fall in the range of 10 to 20 h^{-1}. However, due to the uncertainties and inconsistencies in our current knowledge of both BOD and DO_{Cr} for plant cell suspension cultures, experimental verification of this information for particular systems of interest is advisable. Since the mechanisms of enhancing $k_l a$ (high impeller RPM and aeration rate) can adversely affect cells, it is desirable to operate relatively close to the minimal OTR of the culture.

Adverse Effects of Power Input to the Bioreactor

Shear Damage Due to Mechanical and Pneumatic Agitation

Problems associated with poor mixing or inadequate mass transfer can be avoided by operating at high power inputs. Therefore, in the absence of detrimental effects of agitation and aeration, the requirements of plant cell culture should be easily met. Unfortunately, this is not the case, since mixing and oxygen transfer must be balanced with the detrimental effects of shear. Even though shear rate and shear stress are well-defined fluid dynamic principles, quantification of these parameters is very difficult in the complex geometry of mechanically agitated reactors. When considering disruption, the parameter of interest is typically maximum shear, which is correlated with impeller tip speed.[33,34] The use of tip speed as an indicator of maximum shear rate has some theoretical basis; however, this analogy is based heavily on arguments drawn from use of the Rushton impeller. This impeller is specifically designed to provide a defined high shear region in the tip region to facilitate bubble breakup within gas-liquid reactors (Figure 6). While the maximum linear velocity is still encountered at the tip of an axial flow impeller, the flow patterns are very different, so that the actual shear rates will not likely conform to the same proportionality with tip speed as in the case of Rushton impellers.

It should not be surprising that reactors which were designed for high shear to facilitate high rates of mass transfer are inadequate for the culture of plant cells which are shear sensitive.[10] Figure 7 compares photographs of *Hyoscyamus muticus* cells grown in a shake flask and a bioreactor with a Rushton impeller

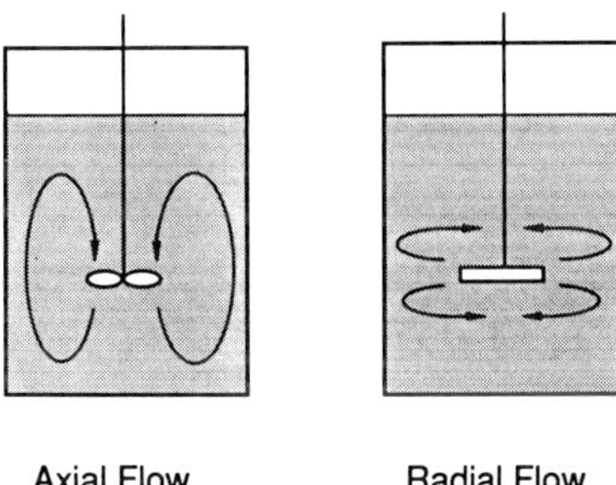

Figure 6. *Flow patterns for an axial and a radial (Rushton) flow impeller.*

at a tip velocity of 42 cm/s. The increase in cell debris in the bioreactor compared to the gyratory shake flask control is quite dramatic. In this particular case, the increase in stress is reflected in the corresponding reduction in growth by approximately 70% in the reactor as compared to the control. Any design and scale-up of a bioreactor for plant cell suspension culture therefore require a quantitative analysis of shear sensitivity. Very few systematic studies have been conducted to evaluate the effect of shear or impeller speed. Figure 8 shows the effect of impeller speed on specific growth rate for various plant cell suspension cultures. The specific growth rate initially shows an increase with increasing impeller speed, probably due to improved mixing and oxygen transfer. However, at higher impeller speeds the growth rate begins to decline, presumably due to increasing shear damage to cells. It is possible to adapt cells to grow under high shear conditions which effectively results in selecting smaller cell sizes.[35] The implications of such an adaptation to secondary metabolite formation must be considered.

Shear damage could also result from sparging. In mammalian cell culture, sparging of gases through the culture has been shown to cause a decrease in cell viability due to shear stress associated with bubbles bursting at the liquid surface.[36] However, direct detrimental effects of sparging, analogous to these, have not been observed for plant cell culture. It can be expected that plant cells will interact less with bubbles due to the presence of cell wall and due to their larger size and tendency to grow in relatively large aggregates. In an attempt to determine if bubbles bursting at the surface could be responsible for cell death, plant cells ejected at the surface of a bubble column were captured for examination. Microscopic examination of these cells did not reveal any gross morphological damage, and there was no observable difference in viability using Evan's blue staining.[37] To further examine cell viability, the 'open bubble' column was sterilized and cells ejected by bubble rupture were 'caught' under aseptic conditions on 'nurse cultures.' Nurse cultures were prepared by wetting sterile filter paper with spent culture medium and placing it on petri dishes containing a solid agar medium. (This procedure provides conditions that will permit individual cell growth; in the absence of a 'conditioned'

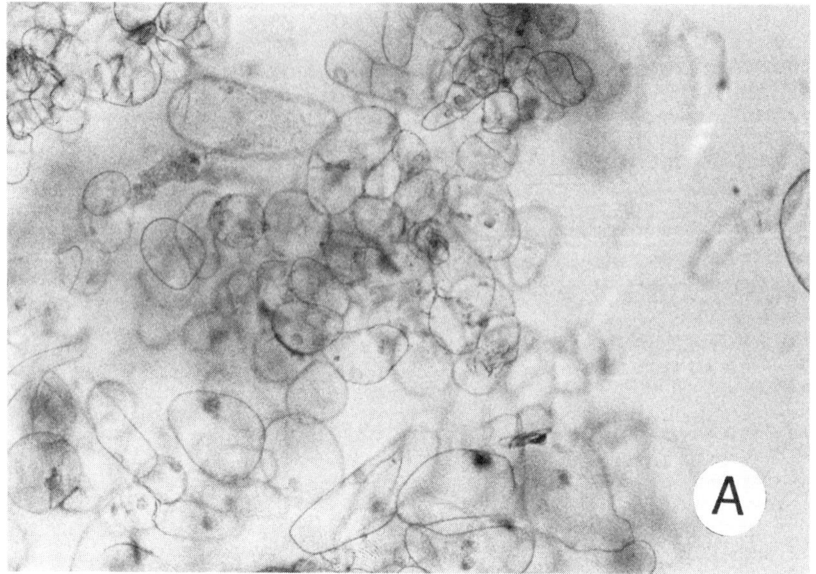

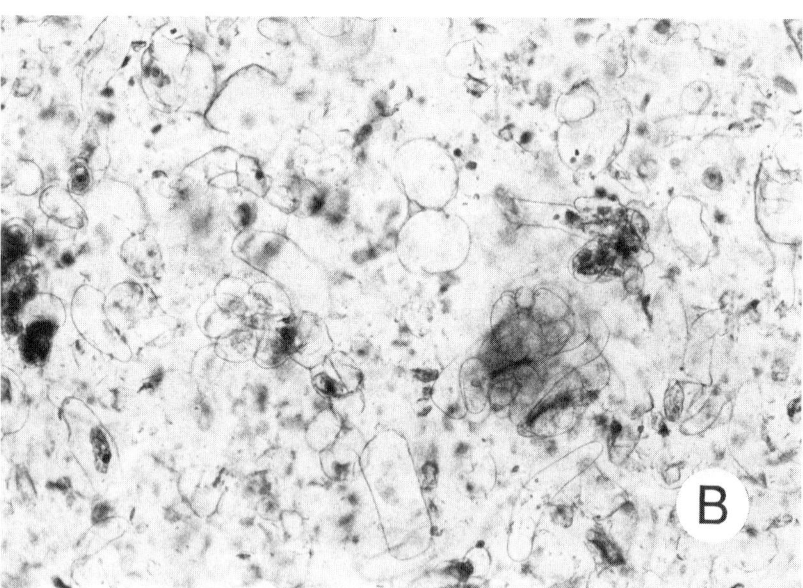

Figure 7. Hyoscyamus muticus *cell suspensions observed under a microscope. (A) Shake flask; (B) 5-l New Brunswick BioFlo III™* bioreactor with Rushton impeller, 100 rpm. (*Registered trademark of New Brunswick Scientific Company, Inc., Edison, NJ.)*

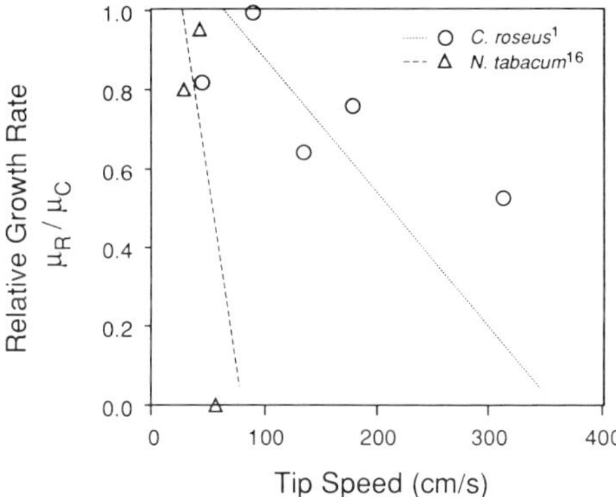

Figure 8. *Effect of impeller speed on growth rate of plant cell suspensions in mechanically agitated vessels. The relative growth rate is expressed as the ratio of the specific growth rate in the reactor (μ_R) to the specific growth rate of controls grown in shake flasks (μ_C).*

medium, all cells would die due to minimum inoculation density problems.) The ejected cells captured in this manner subsequently grew into large clumps of callus. While this clearly demonstrates that the cells are viable, it does not distinguish between 'being alive' and 'being happy'. While the extent of damage imparted by bubbles is clearly not as extensive as for mammalian cells in culture, the same mechanisms may impart damage over the extended periods of tissue growth.

The lack of understanding of both mechanical shear and bubble damage has led to the common practice of using impeller speeds and aeration rates that are 'as low as possible'. The implications of this to design and scalability will be addressed in subsequent sections.

Inhibitory Effects of Aeration

The power input to a vessel is the sum of the mechanical and pneumatic power inputs. Hence, aeration also contributes to the power input to a bioreactor. Although the OTR can be increased by higher gassing rates, high aeration rates have also been found to inhibit culture growth. Hegarty et al.[14] concluded that this inhibition was not due to oxygen toxicity, but resulted instead from the reduction in dissolved carbon dioxide at high air flow rates. Other groups have reported the adverse effects of carbon dioxide stripping on production as well.[38–40] Schaltmann et al.[39] recirculated exhaust gas to obtain a gas regime in their 3-l stirred-tank reactor that was comparable to the shake flask gas regime. This resulted in growth and production in the bioreactor, comparable to that in shaker flasks. Kim et al.[40] report the importance of carbon dioxide as well as ethylene on production of secondary metabolites from *Thalictrum rugosum*

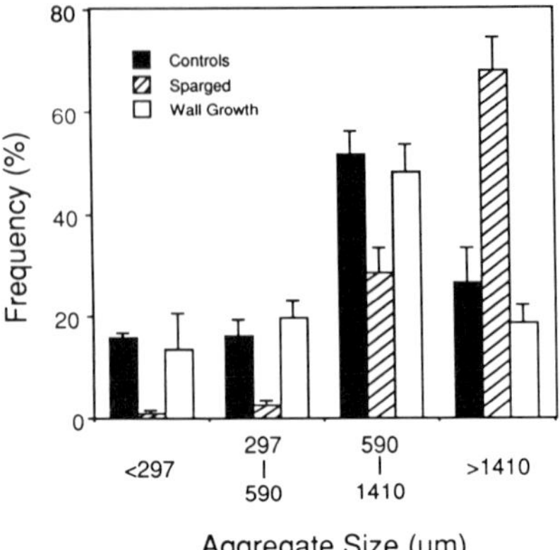

Figure 9. *Comparison of size distribution of* Hyoscyamus muticus *cell suspensions grown in sparged and unsparged (control) 1-l Erlenmeyer flasks (working volume of 250 ml). The size distribution of aggregates occurring as wall growth in the sparged flasks is also represented.*[35]

cell suspension cultures. These studies underscore the importance of gas-phase composition and the resulting liquid-phase concentrations of gaseous compounds. In evaluating studies on the effects of gas-phase composition, it should be kept in mind that it is very likely that cell suspensions of different plant types are likely to behave very differently for the same gas compositions. Ethylene can have both a positive and a negative effect for different plant types;[41,42] therefore, it is likely that there will be no general rule. Other potential indirect effects of CO_2 depletion in air-sparged reactors (such as buffering and oxygen concentration) should be kept in mind as well. Similar to the case of the use of 'low shear' impellers, a practical solution can be based on avoidance as opposed to understanding: oversparging of air should be avoided by providing only as much oxygen as required to maintain aerobic culture conditions.[5,43]

Sparging of air is related to an increase in the observed aggregate size distribution of the cell suspensions as well. Figure 9 compares the size distribution of *Hyoscyamus muticus* cell suspensions grown in 1-l Erlenmeyer flasks, containing 250 ml of B5 media with those of replicated flasks sparged with 0.3 VVM (see Nomenclature) of air. The sparged cultures show a shift towards larger aggregate sizes. In this experiment, the headspace composition of carbon dioxide was observed to be 2 to 4% for the unsparged flasks and only 0.2 to 0.5% for the sparged flasks. This change in gas phase composition supports one of the prevailing hypothesis that changes in gas-phase composition could result in larger aggregate sizes due to general culture stress. Lekcie et al.[18] observed this increase in aggregate size with increase in flow rate as well. Wagner and Vogelmann[7] reported larger aggregate sizes in bubble columns as compared to cultures grown in shaker flasks. Numerous speculations are available to explain the formation of larger aggregates at higher flow rates. In general, the aggregate size increases with an increase in stress which can be attributed in part to changes in dissolved gas composition.

Implications for Bioreactor Design

Reactor design for plant tissue culture can be viewed as achieving a balance between the beneficial and detrimental effects of power input. Conventional mechanically agitated bioreactors for microbial growth are operated at high impeller speeds (2 m/s tip speed) to enhance the oxygen transfer by breaking bubbles by shear. The intense agitation produced by these operational conditions greatly reduces the viability of plant cells. The high aeration rates used in microbial bioreactors (0.5 to 1.0 VVM) are also found to be inhibitory to plant tissue-culture growth. The large power inputs used in microbial culture reflect the high BOD of these cultures. Growth rates of *Escherichia coli* can reach 20 d^{-1} and BOD on the order of 100 mmol/l/h. An analysis of k_la required to support this growth yields a mass transfer coefficient of about 400 h^{-1}. The k_la of a 'typical' mechanically agitated fermenter (0.5 to 1.0 VVM, 100 to 10,000 W/m^3 mechanical power input for a Rushton impeller) is about 50 to 500 h^{-1}, for an airlift reactor (0.5 to 1.0 VVM) the k_la can roughly be between 50 and 500 h^{-1} thereby matching the BOD requirements of microbial culture at the higher levels of power input.

In the previous section, a typical k_la requirement for plant cells was estimated at 10 to 20 h^{-1}. If the relationship between k_la and power input was linear, a 95% reduction in power requirements would be realized for plant cell culture based on the operating conditions for microbial culture. In practice, mass transfer saturates with increased power input; therefore, the reduction in power requirement is not this straightforward. Nonetheless, this simplistic analysis does qualitatively describe the basis upon which operating conditions are chosen for the growth of plant cells in mechanically agitated fermenters. The modifications which are employed in the design of mechanically agitated reactors for plant cell culture involve an overall reduction in the power input, with a simultaneous shift from localized (high shear) power dissipation to distributed (high flow) power dissipation. Power input is reduced in three ways: (1) reduced impeller speed, (2) use of axial flow impellers which have lower power draw, and (3) reduced sparge rate.

Impeller Speed

The power delivered in an agitated vessel, P, is governed by the following equation:

$$P = \left(\frac{\rho}{g}\right) m N_P N^3 D^5 \tag{6}$$

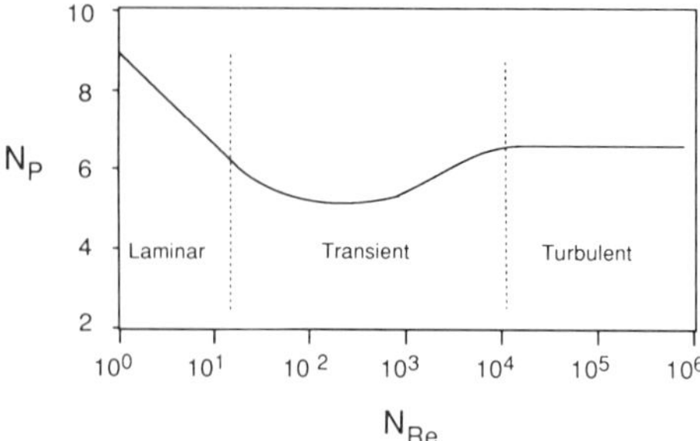

Figure 10. *A typical power number diagram for an impeller which relates the dimensionless power draw of agitation (N_p = power number) to the Reynolds number N_{Re}.*

where m is the number of impellers on the shaft, N_P is the power number for the impeller, N is the impeller speed, and D is the impeller diameter. Clearly, a reduction in impeller speed results in a reduced power input. Large reductions in rotational speed are favored over reductions in impeller diameter because this effectively results in a shift from shear to flow within the reactor. By factoring out impeller tip speed (ND) in Equation 6, it is apparent that the maximum shear rate can be independently manipulated from flow, which is strongly dependent on impeller diameter. A typical impeller diameter is one third the diameter of the tank, while for 'low shear' applications it is common to use impeller diameters approaching one half the tank diameter. A drawback of lower impeller speeds is longer mixing times; however, as described under mixing, fluid-phase homogeneity is less likely to be a problem than plant cell-aggregate suspension. By maintaining flow at the expense of shear, the ability to suspend plant cell aggregates within the system can be enhanced.

Impeller Type

According to Equation 6, the power input can be reduced by reducing the power number. The power number is defined by the power curve of the impeller which becomes constant in the turbulent flow regime, i.e., at high rotation speeds. A typical power number diagram for an impeller is shown in Figure 10. In Table 3, the turbulent power number is given for various types of impellers. It is clear from Table 3 and Equation 6 that, at comparable operating conditions, the power requirement per unit volume of reactor will be the highest for a Rushton impeller. Therefore, the use of an axial flow impeller can reduce the stress caused by high power input. Probably more important than power draw, the use of axial flow 'low shear' impellers results in a shift in the mechanism for power dissipation within the reactor. The power imparted to an impeller is distributed between shear and flow (Q); in traditional agitated

Table 3 *Commonly used impellers and their power numbers in the turbulent region (at $N_{Re} = 10000$)*

Impeller	Power number	Ref.
Rushton, 6 blade	5.5–6.5	
Rushton, 12 blade	8.0–9.0	44
Rushton, 18 blade	9.0–10.0	
Turbine, 45°, 4 blade	1.5–2.3	45
Curved turbine, 6 blade	2.0–4.0	46
Paddle	1.0–3.0	45
Paddle, 45° blade	0.5–2.0	47
Propeller	0.1–1.0	47
Helical	0.8	48
ABEC	0.87	49
Anchor	0.8	48
MIG	0.2–0.4	47

vessel design, shear is often termed 'head' (H) in reference to the high velocity (and pressure) gradients produced by shear.

$$P \alpha Q \cdot H \tag{7}$$

The extreme case for shear is exemplified by a homogenization impeller which produces extremely high velocity gradients and virtually no flow. On the other extreme, 'anchors' and helical impellers produce large amounts of flow but very little shear. In either case, the energy is dissipated within the vessel, but in the case of flow-inducing (low shear) impellers, the viscous energy dissipation is more uniformly distributed throughout the vessel in the wakes of the circulation flow fields. Both axial flow and radial flow (Rushton) impellers are relatively high flow, but the Rushton has a higher shear rate at the impeller tips. As a result, axial flow impellers have proven to be very successful for plant cell culture. Similar to the discussion of the use of low RPM and large impellers, the shift from shear to flow enhances the suspension capacity within the reactor, which becomes important as the overall power is reduced.

Figure 11 compares the growth of *Hyoscyamus muticus* cells grown in a high shear environment (Rushton impeller at 100 cm/s tip speed) to those grown in a low shear environment (ABEC low shear impeller at 40 cm/s tip speed). The culture grown in a high shear environment had almost no accumulation of biomass after 14 d, whereas the culture grown in the low shear environment grew to a fresh weight tissue density of 350 g/l. Thus, the reduction of shear and the accompanying reduction in mass transfer are compatible as long as critical oxygen demand of culture is met.

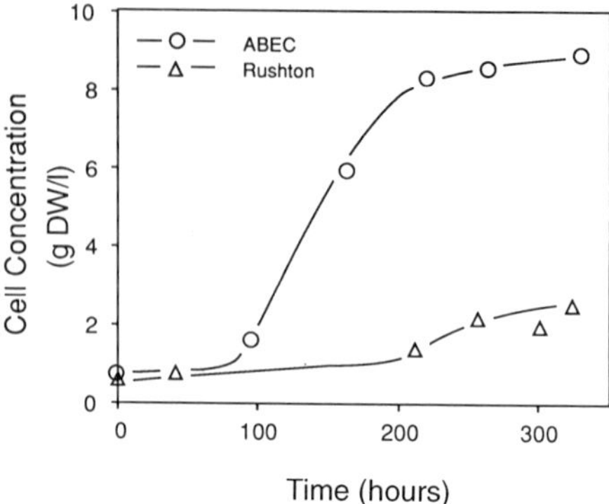

Figure 11. *Comparison of tissue densities of* Hyoscyamus muticus *cell suspensions grown in two different 10-l reactor configurations: high shear (8.9 cm Rushton impeller at 200 rpm in a New Brunswick Micros™ fermenter, orifice sparging, 0.1 VVM) and low shear (13.3 cm ABEC axial flow impeller, 76 rpm in a New Brunswick Scientific Microlift™ bioreactor, 2 μm porous metal sparger, 0.1 VVM).*

Aeration Rate

The relatively low BOD of plant cells also reduces the requirement for aeration. Aeration rates of 0.05 to 0.1 VVM are typically used which is an order of magnitude lower than those employed in for microbial fermentations. The rate of mass transfer within a fermenter is related not only to the volume of air introduced but also the interfacial area. As mentioned above, in conventional fermenters the interfacial area is kept large by bubble breakup induced by impeller shear. Since plant cell fermentations are operated at low impeller shear conditions, the impeller has much less control over the overall mass transfer coefficient. To compensate for reduced shear, plant cell reactors can be sparged with fine bubbles instead of the large bubbles typically introduced into conventional fermenters using orifice spargers. The relative contributions of impeller rotation and sparging to the $k_l a$ in a plant cell culture bioreactor are given in Figure 12. In this study, the mass transfer coefficient was measured using nitrogen sparge/dynamic reaeration technique,[25] with and without agitation. At the low gassing rates typically used for plant cell cultures, mass transfer due to sparging contributed 55% (at 0.2 VVM) to 70% (at 0.05 VVM) of observed mass transfer at the small scale (<15 l). In contrast to microbial culture where mass transfer is controlled by impeller speed, the environment of a plant cell suspension bioreactor should be viewed as an agitated airlift reactor.

Table 4 shows the operating conditions and estimated power requirements of numerous bioreactors used to cultivate plant cells in the literature. From this table, it is clear that the mechanical power input is much larger than the power introduced due to gas sparging. The power input for plant cell suspensions is in the range of 10 to 200 W/m^3 which is an order of magnitude lower than the

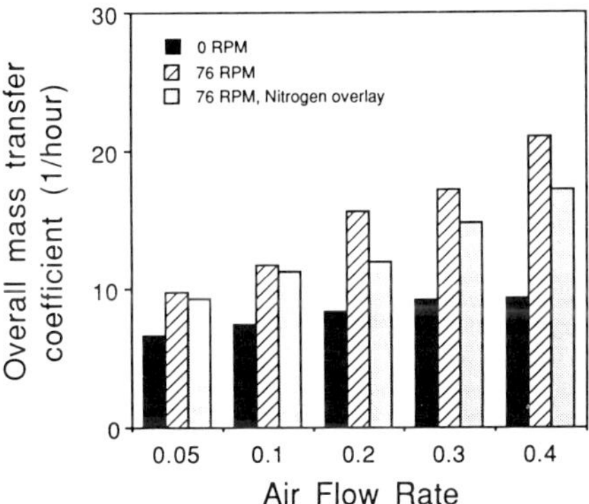

Figure 12. *Overall k_la in a sparged agitated New Brunswick Micros™ bioreactor (10-l working volume). The effect of agitation and surface aeration was studied by conducting the studies at various air flow rates for three conditions: (1) no mechanical agitation (0 rpm), (2) mechanical agitation (76 rpm), (3) mechanical agitation and nitrogen overlay to prevent contribution from surface aeration.*[6]

power input levels for microbial culture. Thus, plant cell suspension cultures require substantially less power input. This is significant in an economic sense because plant cell suspensions require a much longer total fermentation time due to their low growth rates as compared to conventional fermentations. The reduced power requirements for plant cells will contain operating costs over the 2 to 3 week culture period.

Additional Design/Scale-Up Considerations

Contribution of Surface Aeration

In small-scale bioreactors, there is a significant contribution of the culture/ headspace interface to overall oxygen mass transfer rates. Successful scale-up from a lab-scale or pilot-plant scale to an industrial scale requires proper consideration of the effect of surface aeration on overall k_la.[51] By introducing a nitrogen sparge overlay into the vessel, the contribution of surface k_la can be greatly attenuated. At a gassing rate of 0.2 VVM in the 10-l reactor, which was retrofit for plant cell suspension culture as described in Figure 11, the overall mass transfer coefficient was reduced by approximately 20% due to nitrogen overlay (Figure 12). The contribution of the surface is greatly reduced upon scale-up due to the reduction in surface-to-volume ratio. This consideration is particularly important for plant cell cultures in which operational parameters have been determined in small scale cultures at low power input with a close matching of OTR and BOD as prescribed in the overall design strategy for plant cell culture.

Table 4 Operating conditions used for plant cell suspensions grown in stirred tank reactors

Species	Working volume (l)	Impeller	m	D (cm)	RPM	N_{Re}	N_P	P/V[a]	Air sparge rate (VVM)	P/V[b]	μ (day^{-1})	Ref.
Catharanthus roseus	11.0	Helical	1	21	120	88200	—	90	0.1	0.0	0.29	17
Cudriana tricuspidata	5.0	Paddle	1	13.3	60	17689	2[e]	16.6	0.5	10.4	0.11	3
Catharanthus roseus	7.5	Rushton	2	8.5	100	12042	6	32.9	0.05*	1.2	0.32	1
	7.5	Rushton	2	12.0	100	24000	6	184.2	0.05*	1.2	0.25	1
	8.5	Rushton	2	8.5	100	12042	6	29.0	0.01	0.35	0.36	18
Fragaria ananassa	4.0	Flat blade with ribbon	2	6.9[c]	40	5400	6[d]	1.4	0.075	1.78	0	27
Holarrhena antidysentrica	4.0	Turbine	1	6.6[c]	100	5415	6[d]	8.5	0.25	5.5	0.17	50
Hyoscyamus muticus	5.0	Rushton	2	8.25	100	11343	6	42.0	0.4	10.5	0.11	6
	5.0	Axial	2	7.87	150	15484	0.85	16.0	0.4	10.5	0.14	6
	10.0	Rushton	3	8.9	200	26344	6	366.0	0.1	3.3	0.1	6
	10.0	ABEC™	1	13.3	76	22524	0.87	7.5	0.1	3.7	0.2	6

Table 4 (continued) *Operating conditions used for plant cell suspensions grown in stirred tank reactors*

Note: P/V = power per unit volume, W/m^3; m = number of impellers on the impeller shaft; D = impeller diameter; N_{Re} = Reynolds number; N_P = power number; μ = specific growth rate, day^{-1}; * Estimated values.

[a] Power per unit volume calculated based on assumed typical power number.

[b] Power per unit volume based on work done for isothermal expansion of gas.

[c] Diameter of impeller estimated as a third of the tank diameter. Aspect ratio (length/diameter) of tank assumed as one.

[d] Power number for impeller assumed as six. Since Reynolds number at operating conditions is in the transient region, this likely is a slight overestimate.

[e] Power number for a paddle impeller assumed as two.

Bubble Coalescence

Another consideration for scale-up is the observation that bubble coalescence reduces gas-liquid interfacial area which is not be compensated for by impeller redispersion. Small bubbles are very rigid and 'bounce off' one another rather than coalesce. This rigidity results from a high internal pressure due to the combined effects of surface tension and small radius of curvature. As a result, the clouds of small bubbles introduced by fine-bubble spargers are very stable against coalescence. However, once coalescence is initiated, the larger bubbles are more prone to coalescence, which results in an unstable cascade towards larger bubble size and reduced interfacial area per volume of gas in the reactor. In the absence of high shear impellers, coalescence will likely result in a large increase in average bubble size as the depth of the bioreactor increases. This could result in greatly reduced interfacial area for mass transfer upon scale-up.

Aggregate Formation

Aggregate size has been observed to increase with culture stress.[24] In moving from gyratory shake flasks to mechanically agitated and/or sparged reactors, there are numerous mechanisms for elevating culture stress. Not surprisingly, growth of cultures in bioreactors is often associated with an increase in plant cell aggregate size.[7,24] These changes in aggregate size upon scale-up will change requirements for suspension and intra-aggregate nutrient availability. As discussed previously, reactor design can affect aggregate suspension capabilities, and it may also influence nutrient transport within aggregates. Figure 13 shows a cross section of large aggregates collected from a 5-l mechanically agitated reactor and a 5-l concentric cylinder airlift reactor. The aggregate grown in the airlift is hollow indicating that the center of the aggregate died due to nutrient depletion. On the other hand, the aggregate from the stirred-tank reactor is solid and viable to the center. A potential explanation is given in the drawing of Figure 14. In the airlift reactor, the aggregates tended to move with the fluid. As a result, there is little to no relative velocity between the cell aggregate and the surrounding medium and no resulting pressure drop to induce internal flows. The flow in a mechanically agitated vessel is quite chaotic, and the cell aggregate is continuously subjected to a relative flow velocity. The relative velocity between the cell aggregate and surrounding fluid results in a pressure drop which can induce internal flow — analogous to the concepts of effectiveness factor for porous catalysts. This observation would suggest that flow patterns that enhance suspension (such as axial flow impellers) may reduce nutrient availability within aggregates.

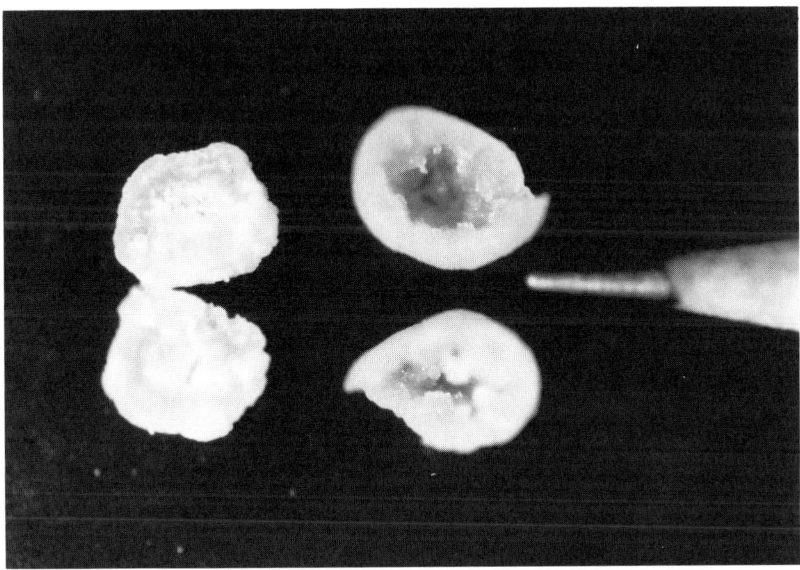

Figure 13. *Cross sections of cell aggregates grown in different reactor configurations. The solid aggregate is from a culture grown in a stirred-tank reactor while the hollow aggregate is from an airlift loop reactor.*

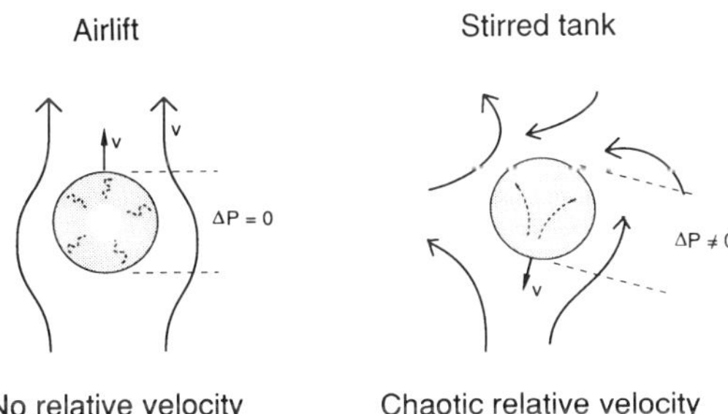

Figure 14. *Flow patterns induced within a cell aggregate due to external flow. A cell aggregate in an airlift loop reactor has no relative velocity with respect to the velocity of the media. A cell aggregate in a stirred tank has a relative velocity with respect to the surrounding media resulting in induced internal flows.*

Fresh Weight to Dry Weight Ratio

The fresh weight to dry weight ratio of plant cell suspension cultures can vary between 5 and 40.[52] By comparison, 1 g of yeast produces only about 3

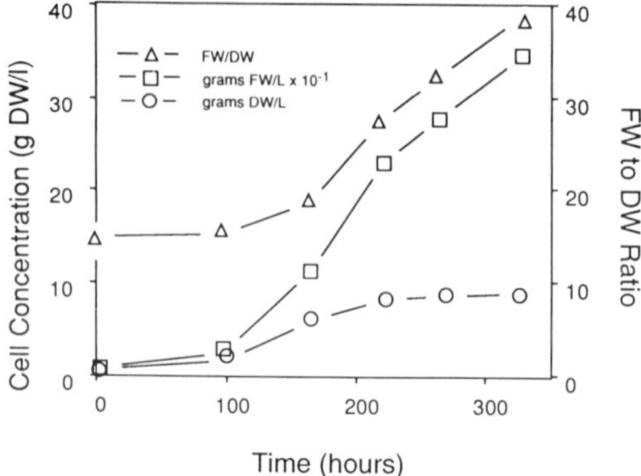

Figure 15. *Fresh weight to dry weight ratio of* Hyoscyamus muticus *cell suspensions cultured in a 10-l for 14 d. The figure also shows the cell concentrations based on fresh and dry weights for the culture.*

g fresh weight. The higher fresh to dry weight ratio for plant cells reflects the large vacuoles and the resulting high cellular water content. The dry weight in a culture is associated with the true biosynthetic biomass and is therefore used as a measure of culture productivity. The fresh weight is more closely related to the volume of the cells. Although the volume fraction of the culture occupied by cells is not a productivity measure, it affects physical parameters which can limit culture growth. The particle volume within a suspension determines its viscosity[53] and rheological characteristics[22] which ultimately determines the ability to mix a suspension of particles.[54] As a result, suspensions with high particle volume cause numerous other problems related to mass and heat transfer.[13] Therefore, as a result of the influence of fresh cell density on reactor transport properties, cellular water content has a strong influence on the maximum achievable dry cell concentration. The wide range of water content observed in plant cell cultures suggests that the maximum achievable dry cell productivities will vary tremendously with different cultures under different growth conditions.[22] This also suggests that efforts to reduce water content would be an effective strategy to increase overall reactor productivity.

As shown in Figure 15, the water content of cells increases markedly during batch cultivation. This observation correlates with the phases on growth in which the lowest water content is associated with the rapidly dividing younger culture, and high water content occurs at the end of culture where cells undergo substantial expansion without division and become highly vacuolated. Drapeau et al.[4] related cellular water content to sugar depletion in the media and the reduction in osmotic pressure in the medium as the culture progressed. It has recently been shown that control of the osmotic environment of the culture can be used to greatly enhance productivity of plant cell cultures. Asali et al.[55] obtained a cell concentration of 40-g DW/l for *Anchusa officinalis* cell suspensions cultured in an osmoticum-controlled perfusion bioreactor. In addition to transient manipulation of water content, media osmoticum can also be used to permanently reduce the water content of cells. In studies of long-term adaptation of tobacco cell suspensions to NaCl, Binzel et al.[56] were able to achieve a 75% reduction in water content without reducing the rate of culture growth. This adaptation reduced the prominent plant cell vacuole to inconspicuous vacuolar strands. These results suggest that substantial reductions in water content can be achieved to enhance the performance of plant cells in reactor systems provided the selection pressure does not adversely affect the ability of the cultures to produce secondary metabolites.

Conclusion

The technology for growing plant cell suspensions at a large scale for the production of chemicals is essentially in place. However, much of the techniques and operating conditions have been arrived at empirically. The resulting lack of understanding of the effects of operational parameters, combined with the tremendous variability in behavior between different cultures and cell lines, still leaves scale-up of plant cell culture in the realm of art as opposed to science. A better understanding of large-scale culture will come with time; however, progress will be slow. The limitations imposed for experimental work are best illustrated with an example: effects of revolutions per minute on culture performance. To carry out a study at five different impeller speeds with only one replication at each speed would require 10 *successful* reactor runs. With a reasonably 'well-behaved' plant cell culture (and an extremely 'well-behaved' graduate student), one could expect to 'turn around' a small scale bioreactor every three weeks. Unsuccessful runs due to contamination, probe and or controller failure, power outages, etc., must be expected, and based on our past experience a completion rate of 75% is not overly pessimistic. Therefore, to complete this rather simple experiment, it would take nearly a year of uninterrupted experimentation by a fully trained researcher familiar with both reactor operation and plant tissue-culture techniques. This clearly shows why those working with plant cell cultures are caught between studying a single culture in depth or superficially examining multiple cultures to provide breadth — both of which are needed.

When the inconvenience of plant cell culture as an experimental system is taken into account, progress on the ability to grow cultures in reactors has been quite rapid. Commercialization has been much less rapid due to unfavorable economics caused by relatively low secondary metabolite titers. Reactor operation can be expected to alter productivity several fold; however, order of magnitude enhancements in secondary metabolite production will require 'brute force' cell line selection (similar to antibiotics from microbial culture). Therefore, reactor design and operational studies facilitate the technical feasibility of large-scale culture but are not likely to play a dominant role in economic feasibility. What is needed are some commercial successes in plant cell culture to drive the technology forward. The development of industrial-scale production will refocus studies on real problems encountered during scalc-up and not simply anticipated problems based on academic speculation. In this manner, industry and academia can work together synergistically to fully utilize the biosynthetic potential of plants for the production of chemicals to benefit mankind.

Nomenclature

μ = Specific growth rate
μ_R = Specific growth rate of cells cultured in a reactor
μ_C = Specific growth rate of the control cultured in a shake flask
υ = kinematic viscosity
ρ = Density of culture broth
ρ_P = Density of cell aggregate
ω_t = Fraction of carbon per gram dry weight of tissue
ω_s = Fraction of carbon per gram sugar (carbon source) in media
ABEC = Low shear impeller design, Elephant Ear Impeller™
BOD = Biological oxygen demand
D = Diameter of impeller
D_P = Diameter of cell aggregate
D_T = Diameter of tank
DO = Dissolved oxygen concentration
DO_{Cr} = Critical dissolved oxygen concentration
DO* = Saturation value of dissolved oxygen concentration in equilibrium with air
DW = Dry weight
FW = Fresh weight
g = Gravitational constant
K = Proportionality constant
m = Number of impellers on the impeller shaft
N = Impeller speed [revolutions per second, rps]
N_{Cr} = Critical impeller speed for suspending cell aggregate of diameter D_p
N_P = Power number
N_{Re} = Reynolds number
OTR = Oxygen transfer rate
P = Power
STR = Stirred tank reactor
RDF = Rotating drum fermenter
V = Volume
VVM = Volumetric of gas per volume of reactor per minute
X = Cell concentration
Y = Cell yield on carbon source

References

1. **Leckie, F., Scragg, A. H., and Cliffe, K. R.,** Effect of impeller design and speed on the large-scale cultivation of suspension cultures of *Catharanthus roseus, Enzyme Microbiol. Technol.,* 13, 801, 1991.

2. **Scragg, A. H., Ashton, S., York, A., Bond, P., Stepan-Sarkissian, G., and Grey, D.,** Growth of *Catharanthus roseus* suspensions for maximum biomass and alkaloid accumulation, *Enzyme Microbiol. Technol.,* 12, 292, 1990.

3. **Tanaka, H.,** Large-scale cultivation of plant cells at high density: a review, *Process Biochemistry,* 8, 106, 1987.

4. **Drapeau, D., Blanch, H. W., and Wilke, C. R.,** Growth kinetics of *Dioscorea deltidea* and *Catharanthus roseus* in batch culture, *Biotechnol. Bioeng.,* 23, 1555, 1986.

5. **Rittershaus, E., Ulrich, J., and Westphal, K.,** Large-scale production of plant-cell cultures, *Int. Assoc. Plant Tissue Cult. News.,* 61, 2, 1990.

6. **Singh, G. and Curtis, W. R.,** unpublished data, 1992.

7. **Wagner, F. and Vogelmann, H.,** Cultivation of plant tissue cultures in bioreactors and formation of secondary metabolites, in *Plant Tissue Culture and Its Bio-Technological Application,* Barz, W., Reinhard, E., and Zenk, M. H., Eds., Springer-Verlag, Berlin, 1977, 245.

8. **Nagochi, M., Matsumoto, T., Hirata, Y., Yamamoto, K., Katsuyama, A., Kato, A., Azechi, S., and Kato, K.,** Improvement of growth rates of plant cell cultures, in *Plant Tissue Culture and Its Bio-Technological Application,* Barz, W., Reinhard, E., and Zenk, M. H., Eds., Springer-Verlag, Berlin, 1977, 85.

9. **Kim, D., Cho, G. H., Pedersen, H., and Chin, C.,** A hybrid bioreactor for high density cultivation of plant cell suspensions, *Appl. Microbiol. Biotechnol.,* 34, 726, 1991.

10. **Mandels, M.,** The culture of plant cells, in *Advances in Biochemical Engineering,* Vol. 2, Ghose, T. K., Fletcher, A., and Blakebrough, N., Eds., Springer-Verlag, New York, 1972, 201.

11. **Smart, N. J. and Fowler, M. W.,** An airlift column bioreactor suitable for large-scale cultivation of plant cell suspensions, *J. Exp. Bot.,* 35, 531, 1984.

12. **Reinhard, E., Kries, W., Bartheln, U., and Helmbold, U.,** Semicontinuous cultivation of *Digitalis lanata* cells: production of β-methyldigitoxin in a 300-l airlift bioreactor, *Biotechnol. Bioeng.,* 34, 502, 1989.

13. **Tanaka, H.,** Technological problems in cultivation of plant cells at high density, *Biotechnol. Bioeng.,* 23, 1203, 1981.

14. **Hegarty, P. K., Smart, N. J., Scragg, A. H., and Fowler, M. W.,** The aeration of *Catharanthus roseus* L. G. Don suspension cultures in airlift bioreactors: the inhibitory effect at high aeration rates on culture growth, *J. Exp. Bot.,* 37, 1911, 1986.

15. **Treat, W. J., Engler, C. R., and Soltes, E. J.,** Culture of photomixotrophic soybean and pine in a modified fermenter using a novel impeller, *Biotechnol. Bioeng.,* 34, 1191, 1989.

16. **Hooker, B. S., Lee, J. M., and An, G.,** Cultivation of plant cells in a stirred vessel: effect of impeller design, *Biotechnol. Bioeng.,* 35, 296, 1990.

17. **Jolicoer, M., Chavarie, C., Carreau, P. J., and Archambault, J.,** Development of a helical-ribbon impeller bioreactor for high-density plant cell suspension culture, *Biotechnol. Bioeng.,* 39, 511, 1992.

18. **Leckie, F., Scragg, A. H., and Cliffe, K. C.,** An investigation into the role of initial k_la on the growth and alkaloid accumulation by cultures of *Catharanthus roseus, Biotechnol. Bioeng.,* 37, 364, 1991.

19. **Shuler, M. L., Pyne, J. W., and Hallsby, G. A.,** Prospects and problems in the large scale production of metabolites from plant cell tissue culture, *J. Am. Oil Chem. Sci.,* 61, 1724, 1984.

20. **Lang, J. A., Yoon, K., and Prenosil, J. E.,** Effect of permeate flux rate on alkaloid production in novel plant cell membrane reactor using *Coffea arabica* cells, *Biotechnol. Progr.,* 6, 447, 1990.

21. **Bramble, J. L., Graves, D. J., and Brodelius, P.,** Plant cell culture using a novel bioreactor: the magnetically stabilized fluidized bed, *Biotechnol. Progr.,* 6, 452, 1990.

22. **Curtis, W. R. and Emery, A. H.,** Plant cell suspension culture theology, *Biotechnol. Bioeng.,* 42, 520, 1993.

23. **Zwietering, T. N.,** Suspending of solid particles in liquid by agitators, *Chem. Eng. Sci.,* 8, 244, 1958.

24. **Kato, A., Asakura, A., Tsuji, K., Ikeda, F., and Iijima, M.,** Biomass production of nitrogen reduced tobacco cells in two-stage continuous culture, *J. Ferm. Technol.,* 58, 373, 1982.

25. **Pirt, S. J.,** *Principles of Microbe and Cell Cultivation,* Blackwell Scientific, Oxford, 1975.

26. **Taticek, R. A., Moo-Young, M., and Legge, R. L.,** Effect of bioreactor configuration on substrate uptake by cell suspension cultures of the plant *Eschscholtzia californica, Appl. Microbiol. Biotechnol.,* 33, 280, 1990.

27. **Hong, Y. C., Labuza, T. P., and Harlander, S. K.,** Growth kinetics of strawberry cell suspension cultures in shake flask, airlift, stirred-jar and roller bottle bioreactors, *Biotechnol. Prog.,* 5, 137, 1989.

28. **van Gulik, W. M., ten Hoopen, H. J. G., and Heijnen, J. J.,** Kinetics and stoichiometry of growth of plant cell cultures of *Catharanthus roseus* and *Nicotiana tabacum* in batch and continuous fermenters, *Biotechnol. Bioeng.,* 40, 863, 1992.

29. **Finn, R. K.,** Agitation and aeration, in *Biochemical and Biological Engineering Science,* Vol. 1, Blakebrough, N., Ed., Academic Press, London, 1967, 69.

30. **Tate, J. L. and Payne, G. F.,** Plant cell growth under different levels of oxygen and carbon dioxide, *Plant Cell Rep.,* 10, 22, 1991.

31. **Pareilleux, A. and Vinas, R.,** Influence of the aeration rate on the growth yield in suspension cultures of *Catharanthus roseus, J. Ferm. Technol.,* 61, 429, 1983.

32. **Asali, E. C., Mutharasan, R., and Humphrey, A. E.,** Use of NAD(P)H-fluorescence for monitoring the response of starved cells of *Catharanthus roseus* in suspension to metabolic perturbations, *J. Biotechnol.,* 23, 83, 1992.

33. **Oldshue, J. Y.,** Fermentation mixing scale-up techniques, *Biotechnol. Bioeng.,* 8, 3, 1966.

34. **Midler, M. and Finn, R. K.,** A model system for evaluating shear in the design of stirred fermenters, *Biotechnol. Bioeng.,* 8, 71, 1966.

35. **Criss, S., Singh, G., and Curtis, W. R.,** unpublished data, 1992.

36. **Allan, E. J., Scragg, A. H., and Pugh, K.,** Cell suspension culture of *Picrasma quassioides:* the development of a rapidly growing shear resistant cell line capable of quasin formation, *J. Plant Physiol.,* 132, 176, 1988.

37. **Handa-Corrigan, A., Emery, A. N., and Spier, R. E.,** Effect of gas-liquid interfaces on the growth of suspended mammalian cells: mechanisms of cell damage by bubbles, *Enzyme Microb. Technol.,* 11, 230, 1989.

38. **Scragg, A. H., Morris, P., Allan, E. J., Bond, P., and Fowler, M. W.,** Effect of scale-up on serpentine formation by *Catharanthus roseus* suspension cultures, *Enzyme Microb. Technol.,* 9, 619, 1987.

39. **Schlatmann, J. E., Nuutila, A. M., van Gulik, W. M., ten Hoopen, H. J. G., Verpoorte, R., and Heijnen, J. J.,** Scaleup of Ajmalicine production by plant cultures of *Catharanthus roseus, Biotechnol. Bioeng.,* 41, 253, 1993.

40. **Kim, D., Pedersen, H., and Chin, C.,** Cultivation of *Thalictrum rugosum* cell suspension in an improved airlift bioreactor: stimulatory effect of carbon dioxide and ethylene on alkaloid production, *Biotechnol. Bioeng.,* 38, 331, 1991.

41. **Pratt, H. K. and Goeschl, J. D.,** Physiological roles of ethylene in plants, *Annu. Rev. Plant Physiol.,* 20, 541, 1969.

42. **Abeles, F. B.,** *Ethylene in Plant Biology,* Academic Press, New York, 1973.

43. **Payne, G. F., Payne, N. N., and Shuler, M. L.,** Bioreactor considerations for secondary metabolite production from plant cell and tissue culture: indole alkaloids from *Catharanthus roseus, Biotechnol. Bioeng.,* 31, 905, 1988.

44. **Rushton, J. H., Costich, E. W., and Everett, H. J.,** Power characteristics of mixing impellers, *Chem. Eng. Prog.,* 46, 395, 1950.

45. **Nienow, A. W. and Miles, D.,** Impeller power numbers in closed vessels, *Ind. Eng. Chem. Process Design Dev.,* 10, 41, 1971.

46. **Bates, R. L., Fondy, P. L., and Corpstein, R. R.,** An examination of some geometric parameters of impeller power, *Ind. Eng. Chem. Process Design Dev.,* 2, 310, 1963.

47. **van't Riet, K. and Tramper, J.,** *Basic Bioreactor Design,* Marcel and Deckker, New York, 1991.

48. **Zlokarnik, M.,** Ruhrtechnik, Band II Verfahrenstechnik, *Ullman, Encyclopedie der Technische Chemie,* Verlag Chemie, Weinheim, 1972, 259.

49. **Slatter, E.,** personal communications, A.B.E. Company, Allentown, PA.

50. **Panda, A. K., Bisaria, V. S., and Mishra, S.,** Alkaloid production by plant cell cultures of *Holarrhena antidysentrica.* II. Effect of precursor feeding and cultivation in stirred tank bioreactor, *Biotechnol. Bioeng.,* 39, 1052, 1992.

51. **Fuchs, R., Ryu, D. D. Y., and Humphrey, A. E.,** Effect of surface aeration on scale-up procedures for fermentation processes, *Ind. Eng. Chem. Process Design Dev.,* 10, 190, 1971.

52. **Curtis, W. R.,** Kinetics of phosphate limited growth of poppy plant suspension cultures, Ph.D. thesis, Department of Chemical Engineering, Purdue University, West Lafeyette, IN, 1988.

53. **Einstein, A.,** Berichtigung zu meiner arbeit: Ein neue bestimmung der molekuldimensionen, *Ann. Physik,* 34, 591, 1911.

54. **Oldshue, J. Y.,** *Fluid Mixing Technology,* McGraw-Hill, New York, 1983.

55. **Su, W. W., Asali, E. C., and Humphrey, A. E.,** Anchusa officinalis: production of rosmarinic acid in perfusion cell cultures, in *Biotechnology in Agriculture and Forestry,* Vol. 21, Bajaj, Y.P.S., Ed., Springer-Verlag, Berlin, Germany, in press.

56. **Binzel, M. L., Hasegawa, P. M., Handa, A. K., and Bressan, R. A.,** Adaptation of tobacco cells to NACl, *Plant Physiol.,* 79, 118, 1985.

Reactor Design for Plant Root Culture

Gurmeet Singh[1] and Wayne R. Curtis[1,2]

[1]Department of Chemical Engineering, The Pennsylvania State University, University Park, Pennsylvania

[2]Biotechnology Institute, The Pennsylvania State University, University Park, Pennsylvania

Introduction

Plant root culture is an alternative production system for plant-derived chemicals that has both advantages and disadvantages associated with a higher level of cellular differentiation. The primary advantage of root cultures is that they reproduce the tremendous biosynthetic capability of roots found in the intact plant.[1,2] The ability to carry out secondary metabolite formation has been directly attributed to a higher level of differentiation as compared to plant cell suspensions.[3] In terms of industrial application as biological catalysts, root cultures have been shown to display stable long-term product profiles which will reduce the amount of selection required to maintain high-producing culture lines. In addition to metabolic stability, roots have also shown extended physical stability under culture conditions. Cultures of *Lithospermum erythrorhizon* have been successfully maintained in a productive state for as long as 220 d.[4] We have subjected nongrowing (phosphate starved) root cultures of *Hyoscyamus muticus* to repeated fungal elicitation for as long as 12 weeks and maintained high levels of sesquiterpene recovery (unpublished

0-8493-8262-9/94/$0.00+$.50

data). This ability of roots to remain viable and productive for extended periods of time, coupled with either secretion or facilitated release of product, emphasizes the role of the tissues as catalysts and de-emphasizes the importance of growth rate in the overall economic viability of the process.

While differentiation provides the advantage of metabolic and physical stability, it also imparts a physical barrier to transport properties within the reactor. Even in the simple gyratory shake flask the root matrix can become fixed within the flask at tissue densities as low as 50 g fresh weight (FW) per liter. Providing adequate mixing becomes problematic even at scales less than 1 l, and problems of stagnation become worse as the tissue concentration is increased in an effort to enhance volumetric productivity. A wide variety of reactor designs and modifications have been used to try to cope with the catalytic root matrix. Essentially all of this work can be characterized as demonstration rather than design. These efforts can be summarized by observing that *at a small scale, plant root cultures can be grown in essentially any reactor configuration* provided that the aeration and mixing conditions are not too extreme. However, the ability to scale-up these reactor systems is much less clear, particularly if high density cultivation is required. Scale-up will be dependent upon the ability to achieve nutrient distribution to roots over very large dimensions which requires analysis of media and gas flow patterns within pilot-scale reactors. To date, there have been less than half a dozen reports on the growth of root cultures at a scale of 10 l or more; therefore, extrapolations to large-scale performance must be placed in a proper perspective. The analysis presented here relies heavily on an analysis of fluid dynamic considerations for root reactor design that has been published elsewhere.[5] This previous analysis will not be reiterated here; instead, this chapter will attempt to synthesize the present status of reactor design for root cultures by combining these preliminary quantitative design principles with the numerous qualitative observations which have been obtained in small-scale reactor studies.

Progress on the Growth of Roots in Reactors

A wide variety of reactor configurations has been utilized for the growth of roots, ranging from conventional stirred tanks to rotating drum bioreactors (Figure 1). As discussed in the preceding chapter for plant cell suspension cultures, the stirred tank would be the simplest configuration for scale-up since it is a proven technology for the 'fermentation' industry. Mechanically agitated fermenters without an immobilization structure have been used, but success has been limited due to problems of shear damage by direct contact with impellers.[6] Taya et al.[7] also reported growth of hairy root cultures of *Beta vulgaris* in a 0.3-l stirred tank reactor and found that the growth rate was much poorer than controls grown in Erlenmeyer flasks. Despite these experimental observations, the possibility of growing roots in submerged suspended culture is attractive — not only because the reactors are commercially available, but

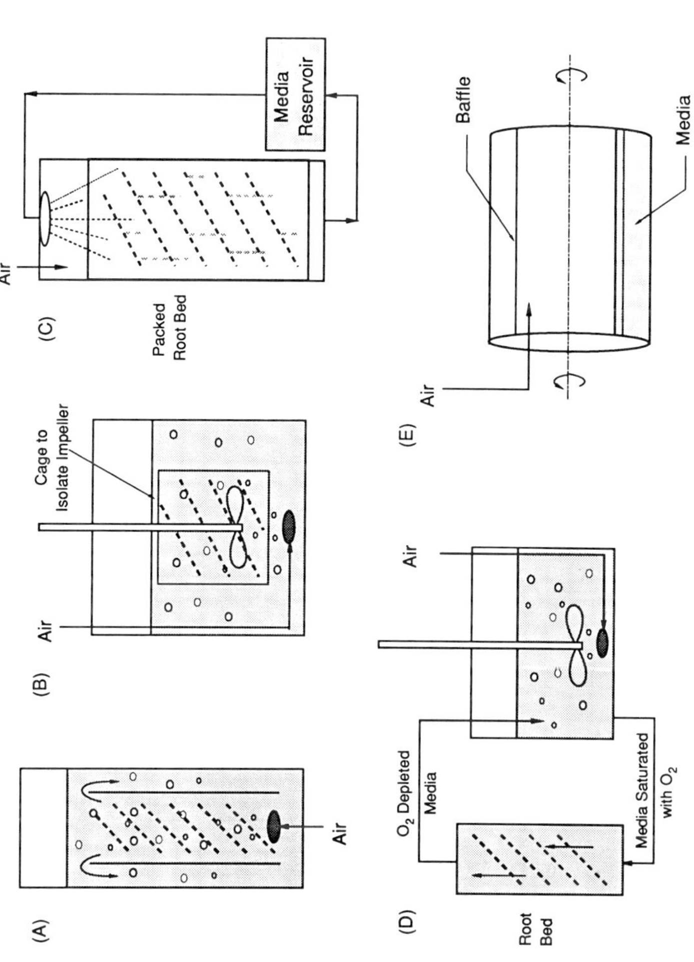

Figure 1. *Reactor types used for culturing plant roots: (A) airlift/bubble column, (B) mechanically agitated, (C) trickle bed, (D) submerged convective flow, and (E) rotating drum bioreactor. The airlift and mechanically agitated reactors can be operated with or without mechanical support for the roots.*

also because root tissue can be easily harvested if grown as a suspension. In the case of secondary metabolites, which are retained intracellularly, the ability to harvest the tissue might outweigh the productivity improvements which could be achieved in alternative higher density 'immobilized' culture systems. The limited studies thus far substantiate the intuition that mechanical agitation can be detrimental for root growth. On the other hand, very little has been done to explore those operational conditions which might possibly support the growth of roots in suspension, such as low RPM or low shear impellers. The remainder of this chapter is based on the premise that high-density culture is an important design objective. While this objective is well-founded, based on the assumption that roots will either secrete or release their secondary metabolites to the surrounding medium, we acknowledge at the onset of this development that there may be some situations where low-density suspended root culture might be more appropriate. In cases where stirred tank fermenters are pursued as potential production systems for root cultures, the principles involved will be similar to those described in the previous chapter for plant cell suspensions.

A modification to the conventional stirred tank which has been used for root cultures is the inclusion of a support matrix to 'immobilize' the roots. While this modification is simple in form, it introduces a large difference in operation. Instead of moving with the flow in the reactor, the roots become stationary and flow takes place through the root matrix. In addition, the 'immobilized' root tissue now grows in a manner which is isolated from the mechanical damage of the impeller. Isolated impeller reactors have been used in various forms and have resulted in good growth performance.[7–9] However, as with other reactor designs, application of isolated impeller systems to reactor volumes greater than 25 l has not been attempted. It can be anticipated that there will be numerous problems associated with providing sufficient mixing in such systems upon scale-up. Attempts have been made to immobilize root tissue in alginate beads to avoid impeller contact and provide better mixing and distribution throughout the vessel; however, this technique has limitations since the root tissue grew right through the beads.[10]

Air-sparged systems without mechanical agitation have also been used successfully for culturing hairy roots. Rhodes et al.[11] used air-sparged systems to grow hairy root cultures of *Nicotiana rustica* to densities of 8 g DW/l. Taya et al.[7] found that growth of hairy roots in airlift reactors increased dramatically if the tissue was immobilized. However, in most cases where comparisons are made between sparged reactor configurations and control gyratory shake flasks, the sparged reactor systems display reduced rates of growth.[12,13] Problems such as excessive gas-phase channeling, blockage of liquid flow patterns, and nonuniform distribution of root tissue have been noted by those who have used these reactors.[10,13,14] Problems associated with sparging alone will also be present in other reactor systems that employ sparging, such as the stirred tank; therefore, it is often difficult to discriminate between possible reasons for poor reactor performance. Another factor which complicates interpretation is the range of aeration rates reported in literature which vary by almost an order of magnitude.

There are numerous other reactor configurations and modifications that have been successfully implemented for the growth of plant root cultures. A notable design that is substantially different from the conventional sparged or agitated reactor systems is the trickle-bed reactor. In this reactor system, the media is recirculated from a reservoir and sprayed over the fixed bed of roots and permitted to flow down through the root bed. The spray can be either a fine mist[21,22] or a coarser spray.[20] Both Wilson et al.[6] and Ramakrishnan and Curtis[19] report biomass densities of more than 10 g DW/l for batch operations in the trickle-bed mode. This reactor configuration has promise in terms of scale-up since the flow patterns are strongly influenced by gravity, which acts uniformly over the bed in contrast to localized power input from mechanical agitation. Analogous to gas-phase channeling encountered in the bubble column, the trickle-bed reactor is subject to channeling of the liquid phase as it flows down over the root matrix. The performance of these reactors will therefore be dependent on the ability to provide sufficient wetting of the root matrix for nutrient availability.

Several reactors have been designed to try to provide both wetting and gas-contacting. Operation of a packed bed of roots with repeated fill and drain cycles is one such approach.[7,18] While this solves the wetting problem at a small scale, the fill and drain times required for larger reactor systems would greatly change the environmental conditions for growth throughout the culture period. With extended fill times, a large fraction of the culture would experience growth conditions that are analogous to the submerged 'plug flow' reactor. Ironically, a simple plug-flow reactor configuration is about the only reactor which has not been used for the growth of roots — most likely due to the technical difficulties of maintaining sterility in the convective flow loop for extended periods of time. A reactor configuration which operates on principles similar to a fill and drain reactor is the rotating drum.[15] In this configuration, the 'filling' takes place as the tissue rotates below the media surface; draining occurs as the tissue rotates up out of the media. This configuration minimizes the problems associated with the timing of the fill and drain cycle, but may have some limitations in scale.

Almost all the studies discussed briefly above and comprehensively tabulated in Table 1 can be viewed as demonstrations of the ability to grow root cultures in bioreactors. Very little has been done to characterize the principles underlying the success and limitations of the various reactor designs. This lack of fundamental information, combined with the tremendous variety of bioreactors and operating conditions used, makes it very difficult to discuss reactor design in the context of specific reactor geometries. In the subsequent section, we develop an alternative framework for discussion of reactor design principles and draw on examples from this heterogeneous collection of data to clarify important concepts.

Table 1 Review of growth of root cultures in reactors

Reactor	Vol. (l)	Comments	Root culture	Time (d)	Inoculum (g DW/l)	Final tissue concentration (g DW/l)	Ref.
STR	1.0		*Tagetes patula*	28	0.05*	9.3	10
	1.0		*Atropa belladonna*	15	0.03	3.26	8
	1.0		*C. sepium*	15	0.03	11.9	8
	0.3		*Armoracia rusticana*	25	0.3–0.5	4.8	7
Isolated impeller	20		*C. sepium*	15	0.03	6.3	8
	1.0		*Daucus cartota*	30	0.2	10	15
	1.0		*Beta vulgaris*	20	1.0	13	16
	12.0	Batch	*Datura stramonium*	35	—	9.7	9
	12.0	Continuous	*Datura stramonium*	37	—	20.6	9
	10.0		*Datura stramonium*	28	—	9.8	6
Bubble column	1.0		*Catharanthus roseus*	26	0.8*	13	17
	2.5		*Atropa bella donna*	26	0.04	1.3	12
	0.8	Continuous	*Datura stramonium*	60	0.1*	40	6
	0.8	20 d batch + 45 d cont. SS cage	*Nicotiana rustica*	65	0.4	27	6
	0.88	13 d batch + 16 d cont. SS cage	*Nicotiana rustica*	29	0.03	8.18	11
	9.0	Draft tube	*T. foenum-graceum*	30	0.03*	—	14
	9.0	Draft tube	*T. foenum-graceum*	60	0.03*	2.8*	14

Table 1 (continued) *Review of growth of root cultures in reactors*

Reactor	Vol. (l)	Comments	Root culture	Time (d)	Inoculum (g DW/l)	Final tissue concentration (g DW/l)	Ref.
	9.0	Nylon mesh	*T. foenum-graceum*	30	0.25*	7.5*	14
	9.0	Nylon mesh	*T. foenum-graceum*	60	0.25*	—	10
	1.0	Immobilized in alginate beads	*Tagetespatula*	28	1.0	6.8	10
	2.0	Continuous	*L. erythrorhizon*	220	0.1	—	4
	0.3	Free	*A. rusticana*	25	0.3–0.5	2.8	7
	0.3	Immobilized	*A. rusticana*	31	0.3–0.5	2.8	7
	0.25	Immobilized	*Hyoscyamus muticus*	18	0.11	3.2	13
Ebb and flow	2.0	Immobilized	*Hyoscyamus muticus*	18	0.22	11.7	18
	0.3	Immobilized	*A. rusticana*	34	0.3–0.5	4.7	7
Trickle bed	15.0		*Hyoscyamus muticus*	28	0.22	11.9	19
	2		*Hyoscyamus muticus*	30	—	—	20
	2.0		*Datura stramonium*	25	—	10.3	6
	0.3		*A. rusticana*	48	0.3–0.5	6.8	7
Nutrient mist	1.4		*Beta vulgaris*	7	0.43[a]	1.3[a]	21
	1.4		*Carthamus tinctorius*	7	0.52	1.5[a]	21
RDF	1.0	Polyurethane foam	*Daucus carota*	30	0.2	10	15

[a] Tissue concentration calculated based on 75% of culture space.

Note: * Estimated from fresh weight by assuming FW/DW = 20.0; SS = stainless steel; STR = stirred tank reactor; RDF = rotating drum fermenter.

Bioreactor Design

The principles for design of reactors for plant root cultures are the same as for plant suspensions: there must be sufficient mixing and mass transfer to provide adequate nutrient availability under operating conditions which are not detrimental to growth. In practice, application of these principles is quite different due to the physical structure of roots. Growth of root cultures in conventional stirred-tank fermenter systems is possible, but the tissue concentration attainable is limited by the ability to maintain flow and obvious problems of mechanical disruption of tissue due to mechanical impact by impellers. The combination of the unique structure of roots and limited suitability of conventional fermentation systems has spawned a wide range of proposed reactor systems discussed in the previous section.

Unlike cell suspension cultures where solid cells are freely suspended in the liquid medium, roots are either fixed on a support upon inoculation or are stationary by virtue of the entwined root matrix formed during growth. In a sense, the root reactor can be treated as a fixed bed of roots with medium circulation providing the necessary nutrients and oxygen. Under optimal conditions, the roots should be uniformly distributed within the reactor, and the medium should have access to all parts of the root matrix. Due to the branching nature of roots, the tissue forms an interlocked matrix, and at concentrations as low as 50 g FW/l, the root mass will become 'fixed' in place within the reactor. While such a biotic phase density might be considered reasonably high for microbial culture, this represents less than 5 g dry weight (DW) per liter due to the relatively high water content of plant tissue. This interlocked bed of root tissue exhibits a large resistance to flow. Since the saturation dissolved oxygen concentration is 0.25 mmol/l, and the biological oxygen demand (BOD) of plant tissues is about 1 to 5 mmol/l/h, the oxygen in the media will be depleted in a matter of minutes if it is not replenished. Therefore, regions of the reactor which experience stagnation or poor mass transfer will become exhausted of available oxygen.

Since oxygen is the only nutrient that is not available in the medium in stoichiometric excess, the effects of scale-up on the availability of oxygen must be assessed. For the purpose of classification, the delivery of oxygen to the tissue can be characterized by three different *operational* modes of transport (Figure 2). In a gas-sparged reactor, the oxygen is delivered by local transfer of oxygen from gas bubbles which are rising through the reactor. In reactor systems that incorporate external pumping (such as a packed bed) or internal pumping (such as the isolated impeller), bulk convective flow of the medium can be the dominant mechanism of oxygen transport, particularly to dense regions of root tissue which exclude bubbles. The final mechanism of oxygen delivery is exemplified in the trickle-bed reactor where the medium is sprayed over a column and intimate gas-phase contacting takes place throughout the reactor. In a 'real' reactor, there will be overlap in mechanisms of oxygen delivery; nonetheless, division of oxygen delivery into these operational modes

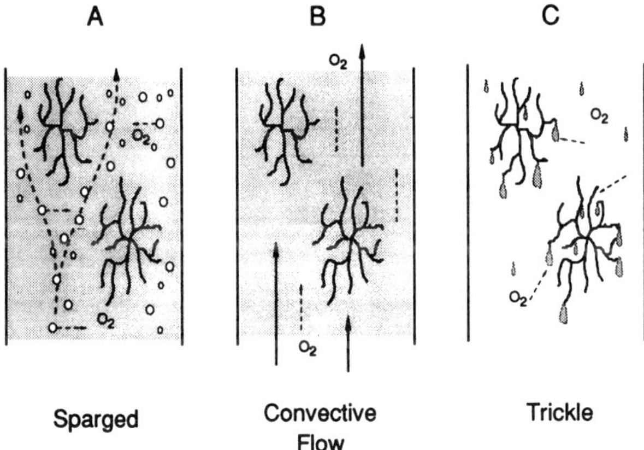

Figure 2. *Operational modes of oxygen transport to root tissue within a reactor: (A) oxygen transfer by gas bubbles rising through the reactor, (B) oxygen transfer by bulk convective flow of liquid, and (C) oxygen transfer by intimate gas phase contacting.*

provides a convenient framework to organize discussion of reactor design principles for fixed beds of plant roots.

Oxygen Delivery by Aeration

As root tissue accumulates within the reactor, the effectiveness of sparging becomes severely impaired. The root matrix becomes a very efficient net that captures and coalesces the bubbles to form pockets of air throughout the column. A photograph of coalesced bubbles trapped within the root matrix is shown in Figure 3A.[20] These bubbles enlarge and eventually surge up through regions in the bed with the least amount of root tissue. This gas flow pattern results in severe channeling and reduction in both overall oxygen availability (due to reduced interfacial area) and local oxygen availability (due to poor gas dispersion). The regions of the reactor that have the highest density of tissue — and highest demand for oxygen — are the least accessible to bubbles. As a result, the oxygen is least available in the regions of highest need. At extremely high tissue loadings, the entrapped gas can form very large cavities which functionally act as trickle-bed sections within the reactor. The implications of trickle flow will be addressed in a subsequent section.

Kondo et al.[15] made an attempt to correlate the performance of a wide variety of reactor geometries to the overall mass transfer coefficient ($k_l a$). They observed a reasonably good correlation between specific growth rate and overall $k_l a$ at tissue concentrations up to about 5 g DW/l. Deviations above this tissue concentration were attributed to a reduction in the specific oxygen uptake rate in response to the reduced availability of oxygen. A fundamental shortcoming of overall $k_l a$ and similar bulk design parameters is that they inherently assume homogeneity within the reactor which may be achievable in

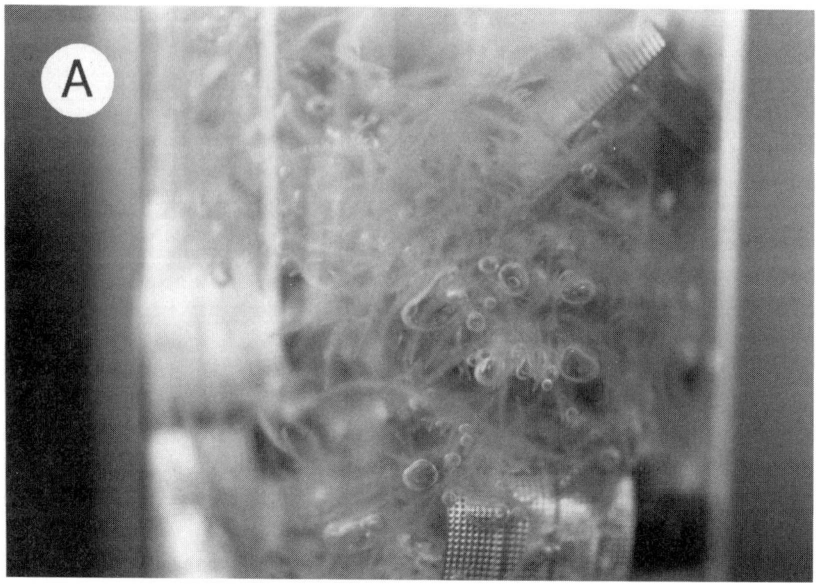

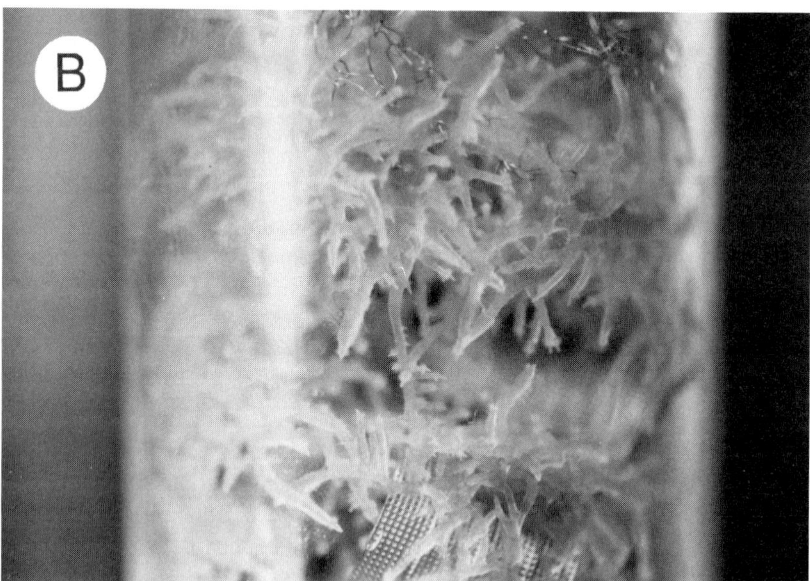

Figure 3. *Comparison of root beds grown as (A) a submerged bed and (B) a trickle bed. These photographs were taken from the same reactor immediately before and after draining.*

very small lab-scale reactors, but will not be valid upon scale-up. Obviously, the assumptions of homogeneous gas and liquid phases will not be valid to provide a sound basis for interpreting an overall k_la even at moderate tissue densities. Due to the nature of root growth in reactors, limitations in operation will necessarily result due to lack of homogeneity. Since the onset of limitations does not preclude continued operation, reactor design should strive to characterize this transition region. A more meaningful measure of distribution of oxygen throughout the reactor space can be obtained by measurements of gas flow patterns. This information could be obtained from gas-phase residence time distributions (RTD), but to date this technique has not been used to characterize root culture reactors.

In addition to the effects of sparging on oxygen availability, sparging may also result in more direct mechanical damage — particularly to root hairs subjected to the rather violent passage of bubble-surges through the bed. To examine the effects of sparging as compared to surface aeration, McKelvey et al.[13] sparged air into flasks of root cultures on a gyratory shaker table. Growth rates with and without sparging were comparable, indicating that oxygen availability was not limiting. However, a partial callusing of tissue and loss of root hairs were noted in the sparged flasks (Figure 4). This change in morphology cannot be exclusively attributed to mechanical damage since the roots also experience a large change in gas-phase composition due to sparging. Stress related to the change in gas composition could also be a factor affecting the morphology. In root cultures grown in gyratory shake flasks, the CO_2 concentration reaches as high as 10% by the time stationary phase is reached. Sparging with air reduces the CO_2 concentration within the reactor due to the drastically reduced residence time of the gas-phase. As described in the preceding chapter for plant cell suspensions, CO_2 may act to alleviate stress associated with ethylene in plant tissue culture; however, in the case of root cultures the data is even more limited than in the case of plant cell suspensions. DiIorio et al.[22] observed higher growth rates in safflower root cultures exposed to air enriched with 1% CO_2, whereas McKelvey et al.[13] reported no enhancement of growth by sparging the cultures of *Hyoscyamus muticus* with air enriched with 10% CO_2 as compared to sparging with air. This comparison exemplifies the situation for interpretation of many potential operational parameters. Data is available for different species, at different experimental levels, and in different reactor configurations — and, not surprisingly, the results are contradictory. Until more information becomes available, it is simply not possible to make generalizations about the sensitivity of reactor performance to these parameters.

Oxygen Delivery by Convective Flow

The problem of inadequate interfacial mass transfer in bubble columns is compounded by the reduction of convection within the vessel caused by the

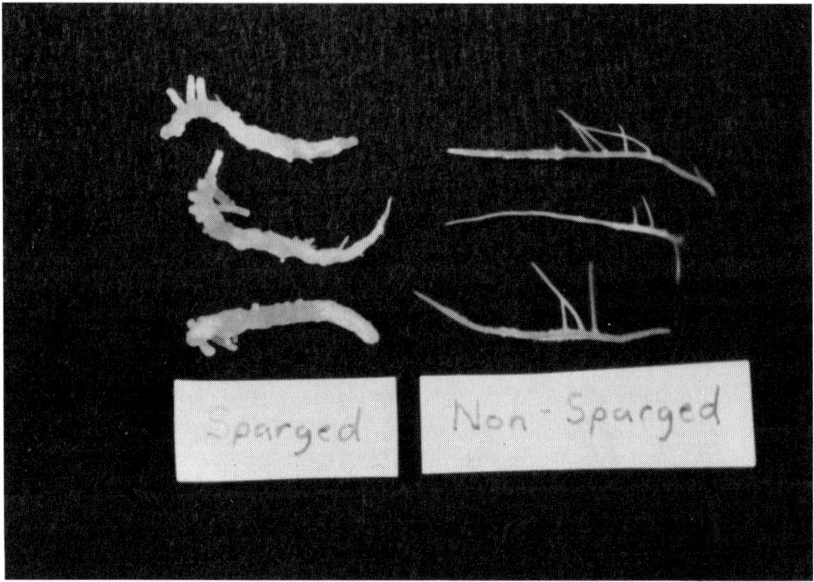

Figure 4. *Effect of gas sparging on morphology of* Hyoscyamus muticus *root cultures. The roots were grown in 250 ml of media in 1-l Erlenmeyer flasks on a gyratory shaker (100 rpm, 2″ stroke). Air was sparged at the rate of 0.16 VVM through a downward facing sintered glass sparger (Corning #39525–20).*

root matrix. To overcome mass transfer limitations which develop in root growth centers, media flow rates on the order of 1 cm/s are required.[23] Reactors such as airlifts can produce circulation rates on the order of 10 cm/s with no flow obstructions; however, it is not likely that this criterion could be met in the presence of roots. Similarly, the extent of mixing in stirred tanks can be assumed to rapidly reduce to zero as one moves away from the impeller in a densely packed bed of roots. The extent of convection in a reactor will be determined by the resistance to flow imposed by the root matrix.

The flow resistance of a packed bed is best characterized from the pressure drop experienced by flow through a tubular reactor. At high tissue loadings, flow through a packed bed of roots can be treated as ideal 'plug flow'. Using tracer techniques, it was recently demonstrated that flow through a packed bed of tobacco roots at 439 g FW/l yielded a residence time distribution which was estimated to be equivalent to 34 CSTRs in series — which is essentially indistinguishable from plug flow.[5] This operational condition would be encountered in high density growth of roots in almost any reactor geometry, including the isolated impeller, draft-tube airlift, and ebb-and-flow reactor configurations. The Ergun equation, given below, provides a semi-empirical description of the relationship between flow-and-pressure drop for a packed bed of cylindrically shaped packing:

$$\frac{\Delta p}{L}=\left[\frac{150}{36}\frac{\mu(1-\varepsilon)^2}{\varepsilon^3}\left(\frac{4}{D_p}\right)^2\right]v+\left[\frac{1.75}{6}\frac{\rho(1-\varepsilon)}{\varepsilon^3}\frac{4}{D_p}\right]v^2 \tag{1}$$

In applying this equation to a packed bed of tobacco roots, it was found that the effective root diameter fell between the diameter of the root axis and the root hair.[5] This indicates that root hairs and branching likely contribute significantly to pressure drop, and pressure drop is likely to be strongly dependent upon the morphology of the specific root culture of interest. Nonetheless, this experimental correlation developed for tobacco provides an example to carry out preliminary design calculations.

As media passes through the root bed, oxygen will be consumed by the tissue. Based on a typical BOD of 4.8 mmol/l/h for plant tissue (estimated from stoichiometry for a cell density of 20 g DW/l and specific growth rate of 0.2 per day), the distance at which oxygen will be completely depleted from the media can be easily calculated for different flow rates and tissue densities. The surface which describes the length of a flow path within a reactor that will result in total oxygen depletion is shown in Figure 5. As expected, depletion increases rapidly with tissue concentration but can be offset by increased flow through the bed. At high tissue concentrations and moderate flow rates, media which is fully oxygenated can become completely depleted over a dimension of several meters. Since complete oxygenation is not likely to take place even in high transfer regions of the reactor, it is clear that the dimensions over which total oxygen depletion can occur are small enough to effect scale-up to commercial size reactors.

The ability to increase flow is achieved by increasing the pressure drop across the root bed. The maximum pressure drop will be dependent upon mechanical strength of the root bed and practical pumping considerations. Using Equation 1, combined with the calculations of oxygen depletion, a surface can be generated which describes the total pressure drop required to move liquid through a packed bed equal to the length at which total oxygen depletion occurs (Figure 6). This surface can be viewed as the minimum pressure drop required to prevent oxygen-depleted conditions over a flow path of specified length and tissue density. Much of this 'operating surface' could not be practically achieved in a plug flow reactor. The pressure driving force for flow in mechanically and pneumatically agitated vessels will be even lower. Not surprisingly, it has been observed that obstruction of the flow path stagnates circulation completely in pneumatically agitated loop reactors.[10,14]

In view of the resistance to flow presented by the root matrix, it is important to realize that oxygen-limited conditions likely exist even in the gyratory shake flasks. Most of the studies conducted comparing the performance of bioreactors for hairy root cultures consider the shake flask as the control. However, the mixing ability of a shake flask is limited, especially when it contains a densely packed root matrix. It is likely that the internal region of the root matrix grown in shake flasks is oxygen-depleted. We have observed that the specific growth

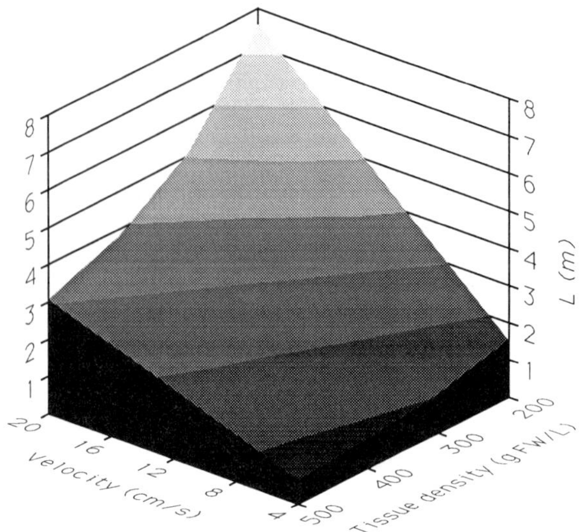

Figure 5. *Length of the root reactor bed that will result in total oxygen depletion as a function of the tissue concentration and flow rate. The calculation is based on a BOD of 4.8 mmol/l/h determined stoichiometrically assuming a tissue density of 20 g DW/l, FW/DW of 20, and specific growth rate of 0.2 per day (refer to Equation 3, Chapter 8).*

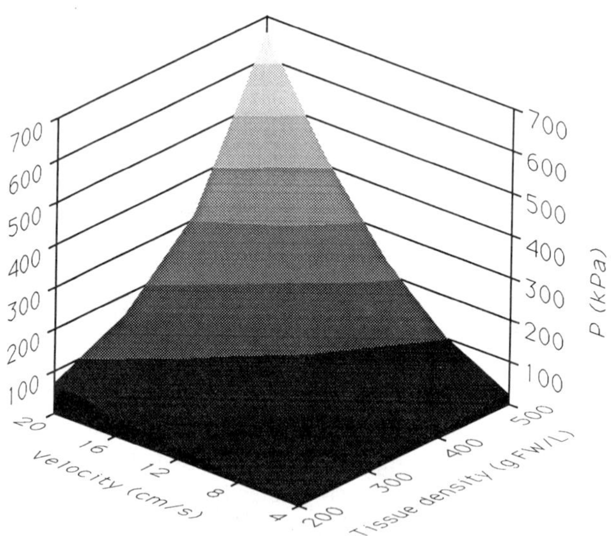

Figure 6. *Total pressure drop required to push media through the length of reactor bed until complete oxygen depletion at specified tissue concentration and media superficial velocity (calculated in Figure 5).*

rate of *Hyoscyamus muticus* root culture is nearly doubled when cultured at lower tissue concentrations or when tissue is pulled apart throughout the culture period to minimized liquid phase, mass transfer limitations.[24] It is therefore likely that the gyratory shake-flask 'benchmark' is an underestimate of the potential for growth if mass-transfer limitations to growth are removed.

Another complication of interpreting scalabilty in submerged reactor systems is that oxygen deprivation does not imply immediate cell death. Plant tissue is capable of surviving anaerobic conditions (anoxia) for extended periods of time. Roots, in particular, have to cope with periods of oxygen deprivation due to flooding conditions in nature. In a preliminary experiment examining the effects of oxygen deprivation, it was found that a 24-h acclimatization to low oxygen concentrations (2% oxygen in the gas phase) permitted sustained viability for nearly a week after switching to a sparged gas of pure nitrogen. On the other hand, an abrupt change from 21 to 0% oxygen resulted in essentially immediate cell death.[25] Since oxygen deprivation will be gradual in a growing root mass, some degree of oxygen deprivation may be tolerable in large-scale reactor systems. This indicates that exhaustion of oxygen may not be a 'clear-cut' design and scale-up criterion.

Oxygen Delivery by Trickle Flow

Trickle-flow contacting provides an alternative operational strategy which permits replenishment of the oxygen in the liquid phase as it flows down through the bed of roots. In the trickle-bed reactor, the medium is recirculated and sprayed over the top of the roots and permitted to drain down through the bed. This type of flow would also be found in fill-drain reactor operations, rotary drum reactors, and densely packed bubble columns which are gas flooded. Since the trickle-bed reactor provides the most defined trickle-flow characteristics, the discussion will be presented in this context. The advantage of trickle-flow operation can be best understood by referring back to Figure 3 which shows a bed of roots growing as a bubble column (Figure 3A) as compared to a trickle bed (Figure 3B). By operating the reactor in the predominantly gas phase, the problem of gas entrainment observed in sparged reactors can be greatly attenuated. Although gas phase RTDs have not been reported for root culture reactors, one can anticipate that the bubble column will be characterized by a very broad RTD, whereas the trickle bed will have a very narrow RTD, indicating that the gas flow pattern in the trickle-bed reactor resembles a plug flow. A plug flow of gas would indicate uniform availability of gas over the entire cross-section of the reactor.

Figure 7 is a photograph of a 2-l trickle bed along with the media circulation loop. The reactor can be operated in both countercurrent and co-current modes. While a countercurrent (up-flow) of gas provides the maximum driving force for mass transfer, in practice a co-current flow (down-flow) of air permits operation to a higher density by eliminating the possibility of flooding.

Figure 7. *A 2-l trickle-bed reactor used for the growth of* H. muticus *root cultures where the recirculated media is sprayed over the root bed and drained into a reservoir. The majority of media is retained in the reservoir resulting in a predominantly gas-phase reactor.*

Ramakrishan and Curtis recently reported the successful operation of a 15-l trickle-bed root reactor for the growth of *Hyoscyamus muticus* at a concentration of 237 g FW/l.[19] An evaluation of liquid flow — including residence time distribution studies — has been reported elsewhere.[5] The findings of this preliminary study are consistent with the general concepts employed in the design of trickle-bed reactors that are outlined here and can be found in more general descriptions elsewhere.[26]

Since the roots are exposed to a predominantly gas phase, oxygen availability only becomes a concern in the context of wetting and liquid flow patterns through the reactor. At moderate to high concentrations (up to 300 g FW/l), a major concern is providing adequate wetting of the tissue and preventing excessive liquid-phase channeling. At higher tissue loadings, regions of the reactor will become sufficiently dense so that they will remain submerged due to liquid entrainment (for submerged sections of the reactor, oxygen must be supplied by the convective flow of media through the region as described in the previous section). The amount of liquid entrained within the bed at any time during reactor operation is referred to as the total liquid holdup. Total holdup is divided into two components: dynamic holdup and stagnant holdup. Dynamic holdup is the fraction of the total holdup that is free flowing and passes quickly through the column. Since it is continuously being exchanged with fresh media, it is not expected to present a limitation for reactor operation based on the relatively low BOD of plant tissue. The stagnant holdup is comprised of the liquid that penetrates the denser regions of the root bed and takes considerably longer to flow through the column. The stagnant holdup is of interest because the oxygen within this component could become locally depleted within the column. Oxygen must diffuse through this stagnant medium surrounding the tissue before it can be assimilated for respiration. A reduction in stagnant holdup minimizes diffusional resistances for oxygen transfer to the tissue. Tracer studies can be used to obtain an estimate of the stagnant holdup. Estimates of dynamic and static holdup can be determined from liquid-phase RTD measured from the time course of inert dye tracer elution from the root bed. We have recently demonstrated the applicability of RTD techniques to root culture bioreactors. Using blue dextran as tracer, Ramakrishnan and Curtis found that only 1.5% of the total bed volume was stagnant in a 15-l trickle-bed reactor operated in a co-current gas-contacting mode.[5] What is now needed is a systematic examination of effects of various operating conditions on liquid holdup and evaluation of corresponding reactor performance.

Due to constraints of sterility and potential problems of tracer toxicity, measurements of static and dynamic holdup are difficult to obtain from an ongoing reactor run. An alternative parameter which can be used in the design of trickle-bed reactors instead of stagnant holdup is the residual holdup. Residual holdup is the liquid retained in a trickle bed after cessation of flow and can be equated to the entrained liquid in a root bed from which the medium has been drained. Some researchers have used media draining as a means of estimating growth within reactors.[9,12] In doing so, they have assumed that the residual holdup is insignificant and that the weight within the reactor represents the mass of root tissue. For most root cultures, this is clearly not valid. We have observed that roots are routinely capable of retaining a mass of media comparable to the fresh weight of tissue within the vessel,[20] and in extreme cases for root cultures with profuse root hairs, the tissue is capable of retaining several times its own weight in media.[5] Given this large variance in medium retention by different root types, one can expect that residual holdup can only provide qualitative information on scale-up; however, the residual holdup provides less information than the stagnant holdup since it is a measure of the liquid holdup under nonflow conditions.

The residual holdup is less than the total holdup during operation, but greater than the stagnant holdup because some of the dynamic holdup is retained in the column due to capillary effects. In fact, under reactor operating conditions, only about one tenth of the residual medium is found to be stagnant.[5] The rest is constantly replaced by fresh media. The part of the residual holdup which is dynamic is the irrigated component of the residual holdup that presents a minimal barrier to convective mass transfer. The remaining stagnant portion of the residual holdup provides substantial resistance to oxygen transfer to the tissue. In the 15-l pilot-scale trickle-bed reactor, it was found that over 86% of the residual holdup in a bed of *Hyoscyamus muticus* roots was constantly being exchanged with fresh medium. This means that although the liquid is entrained, it is easily replaced when additional medium is introduced into the column. Similar to the large variations in residual holdup observed for different root types, the extent of stagnancy within the residual holdup can also be expected to vary considerably for different root types. It is likely, therefore, that although residual holdup is easier to obtain than stagnant holdup, the wide variations observed in root morphologies may limit the use of residual holdup as a predictive design parameter. To overcome this limitation, we are currently developing aseptic RTD techniques are being developed to permit dynamic measurement of stagnant holdup during culture growth.

Knowledge of the liquid residence time distribution does not completely define oxygen mass transfer in a trickle bed. Unlike a purely convective flow reactor, the medium flowing through the trickle-bed reactor can be replenished by intermittent contacting of the gas phase. As a result, the liquid-phase residence times do not define total oxygen depletion as is the case for convective flow. Moreover, when switching from sparged culture to trickle flow, there is a trade-off of gas-phase channeling for liquid-phase channeling. In the case of root cultures, complete wetting of the tissue is not necessary because the roots are capable of transporting nutrients within the root axis. The ability of roots to respond and compensate for nonuniform liquid distribution is shown in Figure 8. This photograph shows the result of growing root cultures in a polycarbonate box in which the medium was slowly recirculated at 0.3 ml/min and dripped onto the center of the growing root mass.[13] It is clear that the root tissue oriented itself to respond to nutrient availability. Under these growth conditions, a number of the roots displayed substantial secondary development suggesting a higher level of vascularization in response to the requirement for nutrient transport along the root axis. (We have never observed this type of development in submerged culture where nutrients are available along the entire root axis.) While these observations suggest that roots are capable of compensating for nonuniform wetting, neither the extent of wetting achieved nor the required level of wetting needed is easily obtained. By growing roots on an inclined plane, McKelvey et al. demonstrated that roots displayed a 50% reduction in growth where nutrient transport was required over a distance of several centimeters.[13] This observation would suggest that a relatively uniform distribution of liquid throughout the reactor is needed. The extent of interaction of the liquid and gas phase must also be assessed. This can only be accom-

Figure 8. *Growth response of* Hyoscyamus muticus *root cultures grown on a support in a 6×6 cm polycarbonate box with a very low flow rate of media dripped at the center.*

plished through techniques such as contact time distributions where the 'tracer' analysis includes measurements of mass transfer between the liquid and gas phase.

Conclusions

In this chapter we have discussed the design of root reactors from the perspective of providing oxygen to cultures at high tissue density. Instead of presenting oxygen transport from a fundamental perspective (i.e., interfacial, convective, and diffusive mass transfer), we have chosen instead to present oxygen transfer in the context of operational modes of delivery of oxygen which correspond to feasible modes of reactor operation for plant root culture. Although this approach is more phenomenological than rigorous, it provides a convenient way of approaching design for these complex, multiphase reactor systems, which are not yet amenable to the 'first principles' approach to design. It is quite clear from this analysis that we are just beginning to scratch the surface on design of reactors for root culture. It should be equally clear that the information needed for design cannot be obtained from small-scale demonstrations of (novel) laboratory reactors. The required data for scale-up can only be obtained from studies focused specifically on obtaining the relevant design parameters.

Nomenclature

ΔP	= Pressure drop
ε	= Packed bed porosity
μ	= Viscosity
ρ	= Density
CSTR	= Continuous stirred-tank reactor
BOD	= Biological oxygen demand
Dp	= Diameter of root axis
DW	= Dry weight of tissue
FW	= Fresh weight of tissue
L	= Length of the reactor bed
RTD	= Residence time distribution
VVM	= Volume of gas per volume of reactor per minute

References

1. **Flores, H. E. and Filner, P.,** Metabolic relationships of putrescine, GABA and alkaloids in cell and root cultures of Solanaceae, in *Primary and Secondary Metabolism of Plant Cell Cultures,* Neumann, K. H., Barz, W., and Reinhard, E., Eds., Springer-Verlag, Berlin, 1985, 174.

2. **Hamill, J. D., Parr, A. J., Rhodes, M. J. C., Robins, R. J., and Walton, N. J.,** New routes to plant secondary products, *Bio/Technology,* 5, 800, 1987.

3. **Flores, H. E., Hoy M. W., and Pickard, J. J.,** Secondary metabolites from root cultures, *Trends Biotechnol.,* 5, 64, 1987.

4. **Shimomura, K., Sudo, H., Saga, H., and Kamada, H.,** Shikonin production and secretion by hairy root cultures of *Lithospermum erythrorhizon, Plant Cell Rep.,* 10, 282, 1991.

5. **Ramakrishnan, D. and Curtis, W. R.,** Fluid dynamic studies on plant root cultures for application to bioreactor design, in *Advances in Plant Biotechnology: Production of Secondary Metabolites,* Furusaki, S. and Ryu, D., Eds., Elsevier, Amsterdam, 1994, 281.

6. **Wilson, P. D. G., Hilton, M. G., Robins, R. J., and Rhodes, M. J. C.,** Fermentation studies of transformed roots, *Bioreactors Biotransformations,* Moody, G. W. and Baker, P. B., Eds., Elsevier, London, 1987, 38.

7. **Taya, M., Yoyama, A., Kondo, O., Kobayashi, T., and Matsui, C.,** Growth characteristics of plant hairy roots and their cultures in bioreactors, *J. Chem. Eng. Jpn.,* 22, 84, 1989.

8. **Jung, G. and Tepfer, D.,** Use of genetic transformation by Ri T-DNA of *Agrobacterium rhizogenes* to stimulate biomass and tropane alkaloid production in *Atropa belladonna* and *Calystegia sepium* roots grown *in vitro, Plant Sci.,* 50, 145, 1987.

9. **Hilton, M. G. and Rhodes, M. J. C.,** Growth and hyoscyamine production of hairy root cultures of *Datura stramonium* in a modified stirred tank reactor, *Appl. Microbiol. Biotechnol.,* 33, 132, 1990.

10. **Buitelaar, R. M., Langenhoff, A. A. M., and Tramper, J.,** Growth and thiophene production by hairy root cultures of *Tagetes patula* in various two-liquid-phase bioreactors, *Enzyme Microb. Technol.,* 13, 487, 1991.

11. **Rhodes, M. J. C., Hilton, M. G., Parr, A. J., Hamill, J. D., and Robins, R. J.,** Nicotine production in hairy root cultures of *Nicotiana rustica:* fermentation and product recovery, *Biotechnol. Lett.,* 8, 415, 1986.

12. **Sharp, J. M. and Doran, P. M.,** Characteristics of growth and tropane alkaloid synthesis in *Atropa belladonna* roots transformed by *Agrobacterium rhizogenes, J. Biotechnol.,* 16, 171, 1990.

13. **McKelvey, S. A., Gehrig, J. A., Hollar, K. A., and Curtis, W. R.,** Growth of plant root cultures in liquid- and gas-dispersed reactor environment, *Biotechnol. Prog.,* 9, 317, 1993.

14. **Rodriguez-Mendiola, M. A., Stafford, A., Cresswell, R., and Arias-Castro, C.,** Bioreactors for growth of plant roots, *Enzyme Microbiol. Technol.,* 13, 697, 1991.

15. **Kondo, O., Honda, H., Taya, M., and Kobayashi, T.,** Comparison of growth properties of carrot hairy root in various bioreactors, *Appl. Microbiol. Biotechnol.,* 32, 291, 1989.

16. **Kino-Oka, M., Hongo, Y., Taya, M., and Tone, S.,** Culture of red beet hairy root in bioreactor and recovery of pigment released from the cells by repeated treatment of oxygen starvation, *J. Chem. Eng. Jpn.,* 25, 490, 1992.

17. **Toivonen, L., Ojala, M., and Kauppinen, V.,** Indole alkaloid production by hairy root cultures of *Catharanthus roseus:* growth kinetics and fermentation, *Biotechnol. Lett.,* 12, 519, 1990.

18. **Cuello, J. L., Walker, P. N., and Curtis, W. R.,** Ebb-and-flow bioreactor for hairy root cultures, Paper 91–7528, Am. Soc. Agric. Eng. Int. Winter Meet., Chicago, IL, December 17 to 20, 1991.

19. **Ramakrishnan, D. and Curtis, W. R.,** Application of residence time distribution studies to plant root culture bioreactors, Plant Tissue Culture II — Session 164, AIChE Annual Meeting, Miami, FL, November 1992.

20. **Flores, H. E. and Curtis, W. R.,** Approaches to understanding and manipulating the biosynthetic potential of plant roots, in *Biochemical Engineering VII: Cellular and Reaction Engineering,* Proc. New York Acad. Sci., Pederson, H., Mutharasan, R., and Di Biasio, D., Eds., New York, 1992.

21. **DiIorio, A. A., Cheetham, R. D., and Weathers, P. J.,** Growth of transformed roots in a nutrient mist bioreactor: reactor performance and evaluation, *Appl. Microbiol. Biotechnol.,* 37, 463, 1992.

22. **DiIorio, A. A., Cheetham, R. D., and Weathers, P. J.,** Carbon dioxide improves the growth of hairy roots cultured on solid medium and in nutrient mists, *Appl. Microbiol. Biotechnol.,* 37, 457, 1992.

23. **Prince, C. L., Bringi, V., and Shuler, M. L.,** Convective mass transfer in large porous biocatalysts: plant organ cultures, *Biotechnol. Prog.,* 7, 195, 1991.

24. **Carvalho, E. and Curtis, W. R.,** unpublished data, 1992.

25. **Singh, G. and Curtis, W. R.,** unpublished data, 1992.

26. **Gianetto, A. and Siveston, P.,** *Multiphase Chemical Reactors: Theory, Design, Scale-up,* Hemisphere, New York, 1986.

Index

A

Aeration
  bubble coalescence, 174, 193
  oxygen delivery, 193
  power input, 165, 167, 168, 172
  rate, 167, 172
  shear damage, 162, 163, 165, 168, 195, 196
  surface, 165, 167, 168, 171, 172, 195
Aggregate formation, 174–175
Agitated reactor design, suspension culture, 152–154
Agitation
  mechanical, *see* Impeller
  pneumatic, *see* Aeration
Agrobacterium tumefaciens
  elicitor, metabolite production, secondary, plant cell culture with fungal elicitor, 25–27
  legume
    acetosyringone, 64
    alfalfa, 63
    antinutritional factors, 69
    atmospheric nitrogen, 69
    cauliflower mosaic virus promoter, 67
    glyphosate, 67
    insect resistance, 69
    lentil, 63
    methionine, 68
    nopaline synthase, 63
    pea, 63
    seed storage protein, 68
    soybean, 63
    transfer-DNA, 61
  legume transformation
    marker transfer, 63–64
    regeneration, 62–63
    reporter gene transfer, 63–64
    transgenic legume production, 64–69
Airlift bioreactor, *see* reactor types
Albinism, haploid, 90
Alcohol dehydrogenase, microprojectile bombardment, 44
Alfalfa
  Agrobacterium tumefaciens and, 63
  Stylosanthes humilis and, 64
Alfalfa mosaic virus, microprojectile bombardment, 45
Alloxane, ginseng production, *Panax ginseng* culture, 7
Androgenesis, 79
Anoxia, 199
Anther, haploid, 78
Anthocyanin biosynthetic gene, microprojectile bombardment, 47
Antinutritional factors, 69
Apparatus, particle bombardment, 38–40
Application, impeller, 153, 172
Artificial seed
  desiccation tolerance, somatic and zygotic embryo, 130–132
    desiccation tolerance, 130–136
    gene expression, 142–143
    LEA protein, 143
    oxidative stress, 141–142
    sugar content changes, 140–141
    water binding fluctuation, 137–140
  haploid, 88
  late-embryogenesis-abundant (LEA) proteins, 143
  micropropagation, 123
Atmospheric nitrogen, 69
Attachment of vector DNA to microprojectiles, 40–41
Automation, micropropagation, 114–116
Auxotroph, haploid, 83

B

Baclofen, tissue-cultured ginseng and, 4
Biochemical studies, 87–88
Biolistic parameters, 41
Biolistics, microprojectile bombardment, 38
Biological oxygen demand
  design criteria, 155, 159, 171, 192
  effect of power input, 169
  experimental, 160
  theoretical, 159, 160, 161, 197, 198
Bioreactor design, 154
Biotechnology applications, haploid
  androgenesis, 79
  biochemical studies, 87–88
  chromosome elimination, 79–80
  culture methods, 81
  factors affecting production, 80–81
    donor plant growing conditions, 80
    genotype, 80
    plant, 80–81
    pollen, developmental stage, 80–81
  future prospects, 88–90
  gametoclonal selection, 84–85
  gene transfer, 85–86

germplasm storage, 88
gynogenesis, 79
mutation, 83–84
physiological studies, 87–88
plant breeding, 81–83
production methods, 78–80
protoplast fusion, 86
selection, 83–84
utilization, 81–88
Biotic elicitor, 24
Blood ethanol concentration, 4, 6
Bombardment, microprojectile
alcohol dehydrogenase, 44
alfalfa mosaic virus, 45
anthocyanin biosynthetic gene, 47
apparatus, 38–40
attachment of vector DNA, 40–41
biolistics, 38
biolistic parameters, 41
cauliflower mosaic virus, 44
electric arc, 38
enhanced regeneration system, 43
expression vector, gene, 44–46
gene delivery, factors influencing, 40–42
gene expression/regulation, 47–49
gene gun, 38
genetic transformation, 43–47
geneticin, 44
hygromycin, 44
introduced genes, 46–47
kanamycin, 44
maize ubiquitin gene, 45
microprojectiles, 40
particle bombardment, 47–49
particle flow gun, 39
particle gun, 38
phytochrome, 48
pit damage, 41
problems, 49–50
process, 38
protoplast, 47
recipient systems, 41–42
rice actin, 45
scutella, 43
selective agent, 43
self-fertile transgenic wheat, 47
shock injury, 41
tobacco mosaic virus, 45
transgenic cells, 43–44
transgenic tissues, 43–44
Box-Wilson method, 12
Bubble coalescence
aeration, 174, 193
reactor design, suspension culture, 174

C

Callusan A, 3
Callusan B, 3
Callusan C, 3
Callusan D, 3
Callusan E, 7
Carbachol, ginseng production, *Panax ginseng* culture, 10
Carbon dioxide, 165, 166, *see also* Aeration; Gas composition
Catharanthus roseus, 28, 29, 31
Cauliflower mosaic virus
microprojectile bombardment, 44
mosaic virus 35S promoter, 67
Cell aggregation
due to aeration, 163, 166
intraparticle mass transfer, 174, 175
sedimentation, 155, 156, 158, 166, 174
stress effects, 166, 174
suspension, critical impeller speed, 156, 157, 158
Cell density effects, viscosity, 178
Cell line selection, 177
Cell suspension, reactor design, 151, 153
Chalcone synthase, 27
Channeling
gas phase, 193, 202
liquid phase, 202
Chromosome doubling, haploid, 81
Chromosome elimination, haploid, biotechnology applications, 79–80
Coalescence, *see* Aeration, bubble coalescence
Computer-aided management, micropropagation, 116–117
Contamination, micropropagation, 117–119
Convective flow, *see* Oxygen, mass transfer
Cost, micropropagation, 113
Culture methods, haploid, biotechnology applications, 81
media composition, 81
pretreatment, 81

D

Defense mechanism, 24–25
Deoxy-glucose, 4
Desiccation tolerance, artificial seed, somatic and zygotic embryo, 130–132
cellular changes during, 136–137
desiccation tolerance variation, 130–132, 133–135
gene expression, 142–143
introduction of desiccation tolerance, 135–136

LEA protein, 143
oxidative stress, 141–142
sugar content changes, 140–141
water binding fluctuation, 137–140
Double haploid, 81

E

Economics, 170, 179, 186
micropropagation, 112–114
Effect of speed, impeller, 158, *see also* Mixing; Shear
Electric arc, microprojectile bombardment, 38
Elicitor, 24–27
Embryogenesis
haploid, 79
Panax ginseng culture, ginseng production from, 9–12
Embryoid, 11
Embryoid culture, 9
Enhanced regeneration system, microprojectile bombardment, 43
Eschscholtzia californica, 28
Ethanol concentration, blood, 4, 6
Ethylene, *see also* Aeration, gas composition
reactor design, suspension culture, 165
Ethylnitrosourea, haploid, 83
Expression vectors, plant gene, 44–46

F

Factors affecting production, haploid, biotechnology applications, 80–81
donor plant growing conditions, 80
genotype, 80
plant, 80–81
pollen, developmental stage, 80–81
Fatty acid, haploid, 82
Fresh weight to dry weight ratio, 175–177, *see also* Water content
Fungal elicitor, 24

G

Gamete recombination, haploid, 82
Gametoclonal selection, 84–85
Gametoclonal variation, 83
Gastric secretion, effects of ginseng on, 3–4, 8
Gastrointestinal propulsion, 7–8
Gene delivery, factors influencing, 40–42
microprojectiles, 40
Gene expression/regulation, 47–49
Gene gun, microprojectile bombardment, 38
Gene transfer, haploid, biotechnology applications, 85–86
Gene transformation, 43–47
Geneticin, microprojectile bombardment, 44
Germplasm storage, 88
Ginseng production, *Panax ginseng* culture, 1–22
alloxane, 7
baclofen, 4
blood ethanol concentration, effect on, 4, 6
blood pressure, 6–9
Box-Wilson method, 12
callusans A, B, C, D, E, 3
carbachol, 10
deoxy-glucose, 4
embryogenesis, 9–12
embryoid culture, 9
gastric secretion, 3
pepsin A activity, 3–4, 8
gastrointestinal propulsion, 7–8
high temperature, 9
histamine, 10
hypoglycemia, 7
intestine, effect on, 3–4
multiple shoots, 9
panaxan A, B, C, D, E, 3
pentagastrin, 10
peripheral blood flow, 6–9
pharmacology, 3
plantlet regeneration, statistical methods, 12–21
root formation, 19–21
saponin, 11
stomach, effect on, 3–4
vagal stimulation, 4
Ginsenosides, 11
Glucose, 4
Glyphosate, 67
Gynogenesis, 79

H

Haploid
albinism, 90
anther, 78
artificial seed, 88
auxotroph, 83
biotechnology applications, 79–81, 87–88
factors affecting production, 80–81
donor plant growing conditions, 80
genotype, 80
plant, 80–81
pollen, developmental stage, 80–81
future prospects, 88–90

gametoclonal selection, 84–85
gene transfer, 85–86
germplasm storage, 88
gynogenesis, 79
mutation, 83–84
physiological studies, 87–88
plant breeding, 81–83
production methods, 78–80
protoplast fusion, 86
selection, 83–84
utilization, 81–88
chromosome doubling, 81
double haploid, 81
embryogenesis, 79
ethylnitrosourea, 83
fatty acid, 82
gamete recombination, 82
gametoclonal variation, 83
germplasm storage, biotechnology applications, 88
gynogenesis, biotechnology applications, 79
herbicide tolerance, 86
Hordeum bulbosum, 79
lipid biosynthesis, 87
microspore, 78
organogenesis, 79
protoplast fusion, 86
recessive traits, 82
Heat shock, 27
Herbicide tolerance, haploid, 86
Histamine, 10
Hold-up, liquid, 201, 202
Hordeum bulbosum, haploid, 79
Hormone deletion, 27
Hygromycin, 44
Hypoglycemic activity, 3, 7

I

Immobilization, 7, 152, 188
Impeller
application, plant cell suspension, 153, 172
effect of speed, 158, *see also* Mixing; Shear
power number, power, 168–169
speed
reactor design, suspension culture, 167–168
for suspension, 156, 158, *see also* Cell aggregation
type, reactor design, suspension culture, 168–169
In situ extraction, 24, 28, 31
Inheritance of introduced genes, 46–47
Insect resistance, 69
Integration of introduced genes, 46–47
Interfacial mass transfer, 159–162
Intestine, small, ginseng production from, 3–4

K

Kanamycin, microprojectile bombardment, 44
K_la, *see* Oxygen, mass transfer

L

Labor, micropropagation, 114
Late-embryogenesis-abundant (LEA) proteins, 143
Legume
Agrobacterium tumefaciens
acetosyringone, 64
alfalfa, 63
antinutritional factors, 69
atmospheric nitrogen, 69
cauliflower mosaic virus promoter, 67
glyphosate, 67
insect resistance, 69
lentil, 63
marker transfer, 63–64
methionine, 68
nopaline synthase, 63
pea, 63
regeneration, 62–63
reporter gene transfer, 63–64
seed storage protein, 68
soybean, 63
susceptibility, 63
transfer-DNA, 61
transgenic legume production, 64–69
current applications, 67
future prospects, 68–69
problems, 64–67
Stylosanthes humilis, 64
alfalfa, 64
soybean, 64
Lentil, *Agrobacterium tumefaciens,* legume, 63
Lipid biosynthesis, haploid, 87
Liquid, hold-up, 201, 202
Lithospermum erythrorhizon, 23, 28, 30
metabolite production, secondary, plant cell culture with fungal elicitor, 23

M

Maize ubiquitin gene, microprojectile bombardment, 45
Mass transfer, *see also* Oxygen, mass transfer

coefficient, suspension culture, reactor design, 159
Metabolite production, secondary, plant cell culture with fungal elicitor, 29
  Catharanthus roseus, 29
  chalcone synthase, 27
  defense mechanism, 24
  elicitor, action of, 25–27
  heat shock protein, 27
  Lithospermum erythrorhizon, 23
  with other stressors, 27–28
  phytoalexin, 24
  plant-microbe interaction, 24–25
  protein phosphorylation, 25
  sanguinarine, 27
  secondary metabolites with fungal elicitation, 28–30
  shikonin, 23
Methionine, *Agrobacterium tumefaciens*, legume, 68
Microprojectile bombardment
  alcohol dehydrogenase, 44
  alfalfa mosaic virus, 45
  anthocyanin biosynthetic gene, 47
  biolistics, 38
  cauliflower mosaic virus, 44
  electric arc, 38
  enhanced regeneration system, 43
  gene gun, 38
  geneticin, 44
  hygromycin, 44
  kanamycin, 44
  maize ubiquitin gene, 45
  particle flow gun, 39
  particle gun, 38
  phytochrome, 48
  transformation, plant cells, 40
Micropropagation
  artificial seed, 123
  automation, 114–116, 115
  computer-aided, 116–117, 117
  contamination, 117–119
    for benefit, 119
    identification, 118
    sources, 117–118
  cost, 113
  cultural innovations, 121–122
  disease-free plant, 117
  economics, 112–114
  future developments, 124
  labor, 114
  long-term storage, 121
  microporous polypropylene, 124
  nutrient mist cultivation, 122
  robotics, 114–116
  somaclonal variation, 119–120
  somatic embryogenesis, 122–124
  vitrification, 120–121
Microspore, haploid, 78
Mixing, 154–158, *see also* Cell aggregation
Morphology, root, 195, 197, *see also* Root hairs
Mutation, haploid, biotechnology applications, 83–84

N

Nitrogen, atmospheric, 69
Nopaline synthase, 63
Nutrient
  mist cultivation, micropropagation, 122
  transport, 202, *see also* Cell aggregation, intraparticle mass transfer; Oxygen, mass transfer

O

Organogenesis, haploid, 79
Osmotic pressure, 27, 176
  water content, 176
OUR, *see* Oxygen uptake rate
Oxidative stress, 141–142
Oxygen
  critical oxygen concentration, 161, 162
  delivery, aeration, 193
  demand, reactor design, suspension culture, 155
  mass transfer
    cell suspension, 159, 161, 167, 170
      design criteria, 155, 161
      head space contribution, 10, 171
      measurement, 159
    roots, 190, 191, 201
      by aeration, 193
      by convective flow, 196
      by trickle flow, 192, 193, 199, 200
  uptake rate (OUR), 159, 161

P

Panax ginseng culture, ginseng production, 1–22
  alloxane, 7
  baclofen, 4
  blood ethanol, 4, 6
  blood pressure, 6–9
  Box-Wilson method, 12
  callusan, A, B, C, D, E, 3

deoxy-glucose, 4
embryogenesis, 9–12
embryoid culture, 9
gastric secretion, 3
pepsin A activity, 3–4, 8
gastrointestinal propulsion, 7–8
ginsenosides, 11
glucose, 4
histamine, 10
hypoglycemia, 3, 7
intestine, effect on, 3–4
multiple shoots, 9
panaxan A, B, C, D, E, 3
peripheral blood flow, 6–9
pharmacology, 3
plantlet regeneration, statistical methods, 12–21
root formation, 19–21
saponin, 11
stomach, effect on, 3–4
vagal stimulation, 4
Panaxan A, B, C, D, E, 3
Particle gun, 38, 39
Pea, 63
Pentagastrin, 10
Pepsin A activity, 3–4, 8
Peripheral blood flow, 6–9
Phenylalanine-ammonia-lyase, 27
Phosphorylation, 25
Physiological studies, biotechnology applications, 87–88
Phytoalexin, 24, 25
Phytochrome, microprojectile bombardment, 48
Picea glauca, 49
Pit damage, microprojectile bombardment, 41
Plant breeding, haploid, biotechnology applications, 81–83
Plant cell culture with fungal elicitor, 29
Catharanthus roseus, 29
chalcone synthase, 27
defense mechanism, 24
elicitor, action of, 25–27
heat shock protein, 27
Lithospermum erythrorhizon, 23
with other stressors, 27–28
phytoalexin, 24
plant-microbe interaction, 24–25
protein phosphorylation, 25
sanguinarine, 27
secondary metabolites, 28–30
shikonin, 23
Plant cells, transformation, bombardment with microprojectiles, 40–42, 46–47
apparatus, 38–40
attachment of vector DNA, 40–41
biolistic parameters, 41
expression vector, gene, 44–46
gene
delivery, factors influencing, 40–42
expression/regulation, 47–49
transformation, 43–47
inheritance of introduced genes, 46–47
integration of introduced genes, 46–47
microprojectiles, 40
particle bombardment, 47–49
problems, 49–50
process, 38
recipient systems, 41
injury to, 41–42
transgenic cells, 43–44
transgenic plant, target tissue for production, 43
transgenic tissues, 43–44
Plant root culture, reactor design
bioreactor design, 192–193
growth in reactor, progress of, 186–191
oxygen delivery by aeration, 193–196
oxygen delivery by convective flow, 195–199
oxygen delivery by trickle flow, 199–203
Plantlet regeneration, statistical methods, 12–21
Plant-microbe interaction, 24–25
Pneumatic, agitation, *see* Aeration
Positive effects of power input, 154–162
Power, *see also* Aeration; Shear
impeller power number, 168, 169
power input, 154, 167
scale-up, 158
Power input
aeration, 165, 167, 168, 172
power, 154
definition, 167
suspension culture
adverse effects of, 162–166
reactor design, 167
Power number, 167–169
Pressure drop, 197, 198
Process, transformation, 38
Production methods, haploid, biotechnology applications, 78–80
Protein
phosphorylation, 25, 26, 27
Protoplast
fusion, haploid, biotechnology applications, 86
microprojectile bombardment, 47

R

Reactor design
  aeration rate, suspension culture, 170–171
  aggregate formation, suspension culture, 174–175
  cell suspension, 151, 153
  fresh weight to dry weight ratio, suspension culture, 175–177
  impeller
    speed, suspension culture, 167–168
    type, suspension culture, 168–169
  interfacial mass transfer, suspension culture, 159–162
  plant root culture
    bioreactor design, 192–193
    growth in reactor, progress of, 186–191
    oxygen delivery by aeration, 193–196
    oxygen delivery by convective flow, 195–199
    oxygen delivery by trickle flow, 199–203
  root culture, 185–206, 186, 187, 190
  suspension culture
    aeration rate, 170–171
    aggregate formation, 174–175
    bubble coalescence, 174
    carbon dioxide, 165, 166
    ethylene, 165
    fresh weight to dry weight ratio, 175–177
    gas-phase composition, 165
    impeller
      speed, critical, 156
      type, 168–169
    interfacial mass transfer, 159–162
    mass transfer coefficient, 159
    mixing, 154–158
    osmotic pressure, 176
    oxygen demand, 155
    oxygen uptake, 159
    power input, 167
      adverse effects of, 162–166
      positive effects of, 154–162
    power number, 167–169
    scale-up, 171–177, 172–174, 177
    sedimentation, 155
    shear, 157
    sparging, 163
    surface aeration, 171–173
    viscosity, 176
Recipient systems, transformation, plant cells, bombardment with microprojectiles, 41
  injury to, 41–42
Residence time distribution (RTD), *see* Hold-up
Reynolds number
  definition of power number, 168
  impeller N_{Re}, application summary, 172
  relation to sedimentation, 156
Rheology, *see* Viscosity
Rice actin, microprojectile bombardment, 45
Robotics, micropropagation, 114–116
Root culture, reactor design, 185–206, 186, 187, 190
Root formation, 19–21
Root hairs, 195, 201

S

Sanguinarine, 27, 30
Saponin, 11
Scale-up
  cell suspension culture, 151
  power, 158
  root culture, 183
  suspension culture, 171–177
Scutella, microprojectile bombardment, 43
Secondary metabolites with fungal elicitation, 28–30
Sedimentation, 155
Seed storage protein, 68
Selection, haploid, biotechnology applications, 83–84
Self-fertile transgenic wheat, microprojectile bombardment, 47
Shear
  damage
    aeration, 162, 163, 165, 168, 195, 196
    cells, 162–164, 165, *see also* Aeration
    roots, 186, 188, 195, 197
    suspension culture, adverse effects of, 162–165
  impeller effects, 167, 168
  maximum shear, 157, 162
  suspension culture, reactor design, 157
Shikonin, 23, 28, 30
Shock injury, microprojectile bombardment, 41
Somaclonal variation, micropropagation, 119–120
Somatic and zygotic embryo,
  cellular changes during, 136–137
  desiccation tolerance variation, 133–135
  development, 130–132
  gene expression, 142–143
  introduction of desiccation tolerance, 135–136

LEA protein, 143
oxidative stress, 141–142
sugar content changes, 140–141
water binding fluctuation, 137–140
Somatic embryogenesis, micropropagation, 122–123
Soybean
Agrobacterium tumefaciens, 63
Stylosanthes humilis, 64
Sparging, *see* Aeration
suspension culture, reactor design, 163
Specific growth rate table, 172
Statistical methods, plantlet regeneration, 12–21
Stomach, ginseng effect on, 3–4
Stylosanthes humilis, 64
Sugar content change, 140–141
Surface, aeration, 165, 167, 168, 171, 172, 195
suspension culture, reactor design, 171–173
Suspension culture
agitated reactor design, 152–154
positive effects of power input, 154–162
power input, adverse effects of, 162–166
reactor design
aeration rate, 170–171
aggregate formation, 174–175
agitated, 152–154
bubble coalescence, 174
carbon dioxide, 165, 166
cell line selection, 177
ethylene, 165
fresh weight to dry weight ratio, 175–177
gas-phase composition, 165
impeller
speed, critical, 156
type, 168–169
interfacial mass transfer, 159–162
mass transfer coefficient, 159
mixing, 154–158
osmotic pressure, 176
oxygen, 155, 159
power input, 167
adverse effects of, 162–166
positive effects of, 154–162
power number, 167–169
scale-up, 171–177, 172–174, 177
sedimentation, 155
shear, 157
sparging, 163
surface aeration, 171–173
viscosity, 176

T

Tobacco mosaic virus, microprojectile bombardment, 45
Transfer-DNA, 61
Transformation, plant cells, bombardment with microprojectiles, 40–42, 46–47
apparatus, 38–40
attachment of vector DNA, 40–41
biolistic parameters, 41
expression vector, gene, 44–46
gene
delivery, factors influencing, 40–42
expression/regulation, 47–49
transformation, 43–47
inheritance of introduced genes, 46–47
integration of introduced genes, 46–47
microprojectiles, 40
particle bombardment, 47–49
problems, 49–50
process, 38
recipient systems, 41
transgenic cells, 43–44
transgenic plant, target tissue for production, 43
transgenic tissues, 43–44
Transgenic legume production, 64–67
Transgenicity
cell transformation, 43–44
plant, target tissue for production, 43
tissue, 43–44
Trickle bed reactor, 187, 189, 191, 194

U

UV radiation, 27

V

Viscosity, *see also* Cell aggregation, sedimentation
cell density effects, 178
of plant cell cultures, 157
Vitrification, micropropagation, 120–121

W

Water binding fluctuation, 137–140
Water content, 176, 179
osmotic pressure effect, 176